Glass-based Materials

Advances in energy, environment and health

Online at: https://doi.org/10.1088/978-0-7503-5904-7

Glass-based Materials

Advances in energy, environment and health

Edited by
Sathish-Kumar Kamaraj
IPN-CICATA Altamira, México

Arun Thirumurugan
University of ATACAMA, Sede Vallenar, Vallenar, Chile

IOP Publishing, Bristol, UK

ISBN 978-0-7503-5904-7 (ebook)
ISBN 978-0-7503-5902-3 (print)
ISBN 978-0-7503-5905-4 (myPrint)
ISBN 978-0-7503-5903-0 (mobi)

DOI 10.1088/978-0-7503-5904-7

Version: 20241001

IOP ebooks

British Library Cataloguing-in-Publication Data: A catalogue record for this book is available from the British Library.

Published by IOP Publishing, wholly owned by The Institute of Physics, London

IOP Publishing, No.2 The Distillery, Glassfields, Avon Street, Bristol, BS2 0GR, UK

US Office: IOP Publishing, Inc., 190 North Independence Mall West, Suite 601, Philadelphia, PA 19106, USA

Contents

Contents

Preface

The investigation of glass materials has spanned numerous centuries, transforming from primitive artistry to state-of-the-art technologies that influence our contemporary society. This compendium explores the historical evolution, present-day developments, and extensive range of glass materials, with an emphasis on their physiochemical characteristics, novel implementations, and encouraging prospects.

Commencing with a comprehensive outline of the physiochemical properties of glass and materials composed of glass ceramics, we establish the groundwork for comprehending their distinctive attributes and multifarious applications. Subsequently, we explore the revolutionary progress of optical fibres within the domain of medicine, illustrating their profound influence on diagnostic procedures and therapeutic approaches in healthcare. The revolutionary applications of bioactive glass in dentistry, where these innovative materials are reshaping the dental care landscape, are then highlighted. Subsequently, we delve into the foundational tenets and pragmatic implementations of glass-based materials concerning health, environment, energy, and the environment, emphasising their critical significance in confronting urgent worldwide issues.

Our research further delves into the utilization of glass-based materials to generate energy in electrochemical systems, thereby revealing the capacity of glass-based technologies to sustainably power the future. The subsequent section of our study is devoted to the fabrication and synthesis of quantum dots for use in glass windows, demonstrating how the convergence of nanotechnology and glass science can be utilized to develop intelligent and energy-efficient architectural solutions.

Subsequently, we explore the technological progressions pertaining to photocatalysts mounted on glass substrates, shedding light on their potential to mitigate indoor air quality issues and combat environmental pollution. We conclude by examining the novel application of glass recycled materials repurposed in reinforced concrete and mortars intended for tropical marine environments. This demonstrates the application of the circular economy and sustainability principles in the field of materials engineering. By conducting this extensive analysis, our objective is to offer valuable perspectives on the complex realm of glass materials and stimulate forthcoming advancements that capitalize on the complete capabilities of these adaptable and essential substances. We hope readers will be delighted with the contents of this book.

Dr Sathish-Kumar Kamaraj, IPN-CICATA Altamira, México.
Dr Arun Thirumurugan, University of ATACAMA,
Sede Vallenar, Vallenar, Chile.

Acknowledgements

Above all, we express gratitude to the Divine for bestowing upon us robust well-being and the invaluable opportunity to complete this book. We would like to extend our heartfelt appreciation to the series editor and advisory board for approving our book and for their support and guidance during the writing process. We wish to convey our appreciation to all the writers and reviewers for their substantial contributions and steadfast support in enabling the creation of this book. We express our sincere gratitude to the numerous publishers and writers who generously granted us permission to utilize their tables and figures.

Sathish-Kumar Kamaraj expresses gratitude to the director of the Centro de Investigación en Ciencia Aplicada y Tecnología Avanzada, Unidad Altamira (CICATA Altamira) and the director general of the Instituto Politécnico Nacional (IPN) for their continuous support and provision of necessary resources to facilitate research endeavours. I appreciate the Secretaria de Investigación y Posgrado (SIP) - IPN for their unwavering support of research and investigation activities, as well as their assistance with project SIP:20231443 y 20240941. The Secretary of Public Education (SEP-Mexico) and the National Council of Humanities, Sciences, and Technologies (CONAHCyT-Mexico) have received additional appreciation. He conveyed his indebtedness to Mrs Mounika Kamaraj and Bbg Aarudhraa for offering support to his family.

Arun Thirumurugan extends his heartfelt gratitude to Dr Justin Joseyphus (NIT-T, India), Professor P V Satyam (IOP, India), Dr Ali Akbari-Fakhrabadi (University of Chile, Chile), and Professor R V Mangala Raja (University of Adolfo Ibanez, Chile) for their kindness and guidance. He also wishes to thank Dr R Udaya Bhaskar, Mauricio J Morel (University of ATACAMA, Chile), Carolina Venegas Abarzúa, Yerko Reyes, and Juan Campos Nazer (Sede Vallenar, University of ATACAMA, Chile) for their invaluable support. Additionally, he acknowledges the support from the management and administration at the University of Atacama, Main Campus in Copiapó, Chile. Arun Thirumurugan gratefully acknowledges the financial support from the Agencia Nacional de Investigación y Desarrollo de Chile (ANID) through the PAI project SA 77210070 and expresses his appreciation to El Centro Tecnológico Ambiental, Vallenar, Chile for their assistance.

Editor biographies

Sathish-Kumar Kamaraj

Dr Sathish-Kumar Kamaraj is a Senior Research Professor Titular C in Instituto Politécnico Nacional (IPN)-Centro de Investigación en Ciencia Aplicada y Tecnología Avanzada, Unidad Altamira, México. He obtained a Doctorate in Nanoscience and Nanotechnology (2010–2014) at The Centre for Research and Advanced Studies of the National Polytechnic Institute (CINVESTAV-IPN), CDMX, Mexico with a scholarship from the General Directorate of International Relations- Secretary of Public Education Mexico (SEP-Mexico). During the academic period, he received the Best Student Competition Award at Battelle's Second International Symposium on Bioremediation and Sustainable Environmental Technologies (2013), in Jacksonville, Florida, USA. His thesis won the best PhD award from the Mexican Hydrogen Society (2014). His passion for sustainable natural systems triggers him to integrate his knowledge of the various fields to address the problems in the areas of energy, environment and health. He developed various working prototypes of microbial fuel cells and plant-based microbial fuel cells. He registered various patents in the Mexican Institute of Industrial Property (IMPI) and technology transfer to the industries. He also has relationships with Government agencies and the private sector for circumstantial decision-making. He served as a guest editor, member of the member of the editorial committee, and reviewer for international journals. He participated as a mentor and gold collaborator for the ReachSci Program of the University of Cambridge.

Arun Thirumurugan

Dr Arun Thirumurugan is an assistant professor at the University of Atacama, Sede Vallenar, Vallenar, Chile. He completed his PhD at the National Institute of Technology (NIT), Tiruchirappalli, India. He has worked as a postdoctoral fellow at the Institute of Physics, India, and the University of Chile, Santiago, Chile. His research interests include the synthesis and surface modification of nanomaterials for energy and environmental applications. He has reviewed over 170 articles for various publishers and has served in editorial roles for several journals, including the Journal of Energy Chemistry, International Journal of Energy Research, Journal of Nanomaterials, Social Impacts, Discover Chemical Engineering, Journal of Physics: Condensed Matter, Optical and Quantum Electronics, and Micromachines. He has more than 120 publications in his credit and edited many books for Springer, Elsevier, Taylor & Francis, and IOP Publishing.

List of contributors

Saravanan Alamelu
Department of Biochemistry and Biotechnology, Faculty of Science, Annamalai University, Chidambaram, India

Sahaya Dennish Babu George
Department of Physics, Chettinad College of Engineering and Technology, Karur, Tamil Nadu, India

Felipe Caballero-Briones
Instituto Politécnico Nacional, Materiales y Tecnologías para Energía, Salud y Medio Ambiente (GESMAT), CICATA Altamira, Mexico

José Luis Cabellos
Universidad Politécnica de Tapachula. Dirección de Investigación y Posgrado. Tapachula, Chiapas, México

Natanael Cuando-Espitia
Universidad de Guanajuato-CONAHCyT, Salamenca, Mexico

Luis Diaz-Ballote
Departamento de Física Aplicada, CINVESTAV IPN, Unidad Mérida, Mexico

Naushad Edayadulla
Department of Chemistry, Vel Tech Rangarajan Dr Sagunthala R&D Institute of Science and Technology, Avadi, Chennai, India

Omar Francisco González Vázquez
Division of Postgraduate Studies and Research. Tecnológico Nacional de México, Aguascalientes Campus, Aguascalientes, Mexico

Juan Hernández-Cordero
Instituto de Investigaciones en Materiales, UNAM, Mexico City, Mexico

Edwin Hoil-Camul
Universidad Politécnica de Tapachula. Dirección de Investigación y Posgrado. Tapachula, Chiapas, México

Nivedha Jayaseelan
Department of Biochemistry and Biotechnology, Faculty of Science, Annamalai University, Chidambaram, India

Sathish-Kumar Kamaraj
Instituto Politécnico Nacional (IPN)-Centro de Investigación en Ciencia Aplicada y Tecnología Avanzada (CICATA), Unidad Altamira, Altamira, Mexico

Gurunathan Karuppasamy
Department of Nanoscience and Technology, Science Campus, Alagappa University, Karaikudi, Tamil Nadu, India

Loganathan Kulanthailvel
Department of Physical Chemistry, School of Chemistry, Madurai Kamaraj University, Madurai, Tamil Nadu, India

Khirbet López-Velázquez
Universidad Politécnica de Tapachula. Dirección de Investigación y Posgrado. Tapachula, Chiapas, México

Swetha Madamala
Department of Chemistry, MVJ College of Engineering, Bangalore, India

Luis Maldonado-López
Departamento de Física Aplicada, CINVESTAV IPN, Unidad Mérida, Mexico

Daniel A May-Arroja
Centro de Investigaciones en Optica A.C., Aguascalientes, Mexico

Pugalehdhi Pachiappan
Department of Biochemistry and Biotechnology, Faculty of Science, Annamalai University, Chidambaram, Tamil Nadu, India

Briska Jifirina Premnath
Department of Biochemistry and Biotechnology, Faculty of Science, Annamalai University, Chidambaram, India

Juan Alberto Ríos-González
Ingeniería en Nanotecnología, Universidad de La Ciénega del Estado de Michoacán de Ocampo, Michoacán, Mexico

José de Jesús Serralta Macias
Division of Postgraduate Studies and Research. Tecnológico Nacional de México, Aguascalientes Campus, Aguascalientes, Mexico

Mireya del Socorro Ovando-Rocha
Instituto Politécnico Nacional, Materiales y Tecnologías para Energía, Salud y Medio Ambiente (GESMAT), CICATA Altamira, Mexico
and
Universidad Tecnológica de Altamira, Química Industrial y Nanotecnología, Materiales y Tecnologías para Energía, Salud y Medio Ambiente (GESMAT), Mexico

Vennila Selvaraj
Department of Nanoscience and Technology, Science Campus, Alagappa University, Karaikudi, Tamil Nadu, India

Kalist Shagirtha
Department of Biochemistry St. Josephs College of Arts and Science, Cuddalore, Tamil Nadu, India

Chandaraj Shanmuga Sundari
Department of Chemistry, Vel Tech Rangarajan Dr Sagunthala R&D Institute of Science and Technology, Avadi, Chennai, India

Ananthakumar Soosaimanickam
R&D Division, Intercomet S.L, Madrid, Spain

Manoj Kumar Srinivasan
Department of Biochemistry and Biotechnology, Faculty of Science, Annamalai University, Chidambaram, Tamil Nadu, India

Rafael Valentin Tolentino-Hernandez
Instituto Politécnico Nacional, Materiales y Tecnologías para Energía, Salud y Medio Ambiente (GESMAT), CICATA Altamira, Mexico

Amado M Velázquez-Benítez
Instituto de Ciencias Aplicadas y Tecnología, Universidad Nacional Autónoma de México, Mexico City, Mexico

Kamalesh Balakumar Venkatesan
Department of Biochemistry and Biotechnology, Faculty of Science, Annamalai University, Chidambaram, India

Chapter 1

Historical development, current progress, and scope of glass materials

Vennila Selvaraj, Gurunathan Karuppasamy and K Loganathan

The current chapter summarizes the state of recent advancements and novel glass material developments. Glass is a non-crystalline substance that is frequently transparent, brittle, and chemically inert. It is widely used for functional, technical, and ornamental purposes. Glass materials are changes in surface properties that have a significant impact on the strength, and random surface defects. In the modern world, glasses are used for a wide range of everyday human needs, from drinking goblets to dressing mirrors, from electric bulbs to communication cables, from window glasses to wine bottles, from decorative vases to the enormous variety of chemical glassware. This chapter covers the historical background of glass materials in the Stone Age dating to recent research and development of glass materials and basic structure formation, properties of glass materials, types of glass materials and their application, and future perspectives of glass materials application.

1.1 Introduction

Since glasses are so prevalent in our daily lives, there is currently a nonchalant attitude towards the group of materials known as glasses that did not previously exist. Glass beads discovered in the tombs and gilded death masks of ancient Pharaohs are proof that early Egyptians thought of glasses as priceless resources. Even earlier cave dwellers used chipped chunks of obsidian, a naturally occurring volcanic glass, to make tools and weapons including scrapers, knives, axes, and spear and arrowheads. For thousands of years, raw materials have been melted and used to create glasses by humans. From at least 7000 B.C. ago, Egyptians used to wear spectacles [1].

Glass is an inorganic, non-crystalline substance that has the appearance of transparency, and we can trace the use of glass back to the stone era using archaeological data. Glass formed by volcanic eruptions was used to make some

doi:10.1088/978-0-7503-5904-7ch1

of the weaponry and implements. Various review articles have assessed the usefulness of smart glasses for research objectives and several kinds of studies have been undertaken on smart glass technologies utilized in various sectors. The appropriateness of smart glasses for medical applications was evaluated by conducting a literature search using specified keywords and search engines. The research was analyzed to assess the technology's functionality, usability, and safety in medical settings, as well as its clinical effectiveness, user acceptance, and ethical implications. After carefully examining the input techniques in the literature relating to commercially available smart glasses, it was possible to analyze the products, technology, sensors, and other elements taken into the research [2, 3].

Significant progress was achieved following industrialization throughout the 1800s. The Chance brothers successfully modified blown cylinder glass's cutting, polishing, and gridding techniques in 1839 to lower breakage and enhance surface quality. The ability to create the enormous number of glass panes needed to build a crystal palace emerged in the 1850s. It took until 1905 for machine-made glass panes to be manufactured, as Emile Foucault managed to extract them straight from the glass melt. By combining many steps of the process into a continuous rolling mill in 1919, Max Bicheroux achieved a crucial breakthrough in the manufacture of glass. The glass melt exited the crucible in segments and went between two cooled rollers to produce a glass ribbon. This method would enable the production of a glass pane measuring 3 by 6 m. The float glass technique was created in the 1950s by the Englishman Alastair Pilkington. It included passing a viscous glass melt over a surface-floating bath of molten tin. Tin was chosen because it has a far greater density than glass and can withstand temperatures between 232 °C and 2270 °C in its liquid physical condition [4]. Presently, floating is the most widely used method, accounting for more than 90% of global production of flat glass. Large production facilities open 24 h a day, seven days a week, produce flat glass. This method involves melting the raw components at 1550 °C, then continually pouring the molten glass over a shallow pool of tin at 1000 °C. Depending on the speed of the rollers, the glass floating on the tin creates a smooth, level surface that is nearly equal in thickness. It then begins to cool to 600 °C before entering the Lehr oven for annealing, where it gradually cools to 100 °C to remove any remaining tension [5].

The cast technique is an additional method for creating flat glass. This method creates glass with the necessary thickness by continually pouring molten glass between metal rollers. Glass with a pattern can be created by engraving the rollers to provide the necessary surface roughness or design. Depending on the design, the glass might have two smooth surfaces, one smooth and one rough, or two textured sides. Furthermore, wired glass may be created by sandwiching a steel wiring mesh between two distinct glass strands. Because wired glass can fuse most glass fragments back together after breaking, it is typically used as fire safety glass [6].

There have been recent developments in composite perovskite glasses, considering the present constraints and difficulties, in terms of the established manufacturing techniques and the ensuing optoelectronic properties. Due to its immense potential to serve as the primary component of the next generation of extremely efficient photovoltaic, organic, and inorganic lead halide, perovskites have recently attracted a significant deal of scientific attention. A broad variety of cutting-edge photonic applications, such as light-emitting diodes, lasers [7], photodetectors, and blacklight displays, have also shown their major importance [8, 9]. Nevertheless, two primary aspects raise doubts about the widespread application and broad commercial utilization notwithstanding this extraordinary progress [4]. Like this, it appears that embedding two-dimensional (2D) materials within the transparency of inorganic oxide glasses is a viable method for creating innovative nano-heterojunctions that allow for the controlled tuning and augmentation of the emission characteristics.

This chapter briefly explains the history, recent development, and analysis of glass materials.

1.2 History of glass materials

The Teutonic word 'Glaza' (which signifies amber) is where the word 'glass' originates. The Mesopotamians from the 5th century BC by coincidence found ash when they fired to melt clay vessels to use for glazing pottery or when copper was smelted, yet the origin of glass production is still unknown. A purposeful glass manufacturing technique has been used to describe the discovery of greenish glass beads in Egypt's early 4th century BC pharaohs' burial chambers. Beginning in the second century BC, rings and miniature figurines made with core-wound processes started to appear. The earliest glass design was created around 669–627 BC on clay tablets [10, 11]. In the second century BC, Syrian artisans developed the blowing iron, which allowed them to create a wide range of thin-walled hollow vessels. Excavations have shown that the public baths in Herculaneum and Pompeii were built with glass for the first time during the Roman era. The installation of these panes might have been done with or without a frame, in a metal or wood surround. The production of items like drinking horns, claw beakers, and master jars as well as a rise in the use of glass in the construction of churches and monasteries occurred during the medieval centuries when this method extended to the northern Alpine areas [4].

1.2.1 First golden ages of glass

The first golden age of glass is referred to as the first four centuries of the Christian era. By the time of the Crusades, glass production had resumed on the island of Murano, close to Venice, where soda lime glass (SLG), often known as crystal, was created.

1.2.2 Second golden age of glass

The earliest factory in North America (the United States) was a glass factory established at Jamestown, Virginia, in 1608, but it collapsed after just a year. A second effort at glassmaking by the colonists of Jamestown was made in 1621, but it was abandoned in 1624 because of an Native American raid in 1622 and a lack of laborers. When Caspar Wistar constructed a glassmaking facility in Salem County, New Jersey, in 1739, the glass industry was once again revived in America. The first significant advancement in commercial glassblowing since the blowpipe was made by Pittsburgh, Pennsylvania's Bakewell, Page & Bakewell Co. in 1820. A method of mechanically pressing heated glass was patented by them. Glass usage and production advancements expanded so quickly after 1890 that they were virtually revolutionary. New, significant specialty glasses first appeared in the late 1900s. Chalcogenide glasses, an infrared-transmitting glass that may be used to produce lenses for night-vision goggles, and transparent glass-ceramics, which are used to make cookware, were among the new specialty glasses [11].

The physics and engineering of glass as a material were considerably better known, and Alastair Pilkington devised a new, ground-breaking manufacturing technique (float glass production) in the late 1950s, with which 90% of flat glass is being made today. Optical fibers were created in the 1970s to be used as 'light pipes' in laser communication systems. Long-distance light transmission is maintained by these pipes' brightness and intensity. The 1970s also saw the development of glass types that can securely store radioactive waste for thousands of years [12, 13] (see figure 1.1). Additive manufacturing (AM), a cutting-edge and ground-breaking technique, generates a lot of research interest in materials processing. Due to its substantial benefits of being able to create glass parts with arbitrary and complicated shapes, it is also progressively being employed in the production of glass. The different types of glass as a scheme and their development through the years 2015–2022 are shown in table 1.1 [14].

1.3 What is glass?

Glass is a classic example of a human creation. The first opaque glass produced by humans was discovered in 8000 BC Egyptian pottery. Glass is appealing to people because of its distinctive qualities, including optical transparency, structural rigidity, compositional flexibility, adaptability for property tailoring, and durability. All of these characteristics make glass appropriate for both conventional and cutting-edge technological purposes [15]. Typical uses include those for windows, containers, lighting, lenses, and hand-made art pieces. Some examples of cutting-edge technical applications include the usage of solar glass, laser glass, bioglass, armor glass, optical communication fibers, etc [4].

1.3.1 Types of glasses

According to its chemical composition, the glass may be broadly categorized into two classes:

- inorganic glasses, and
- organic glasses.

Depending on their source (see figure 1.2) there are two different forms of inorganic glasses: natural and inorganic (such as manufactured or artificial) glasses. While artificial glasses were created by melting the basic ingredients and using their manufacturing expertise, natural glasses were found in volcanic rocks. Again, the categories of ordinary and specialist glasses may be used to categorize synthetic eyewear. Glasses that are found all about us and those we use frequently in our

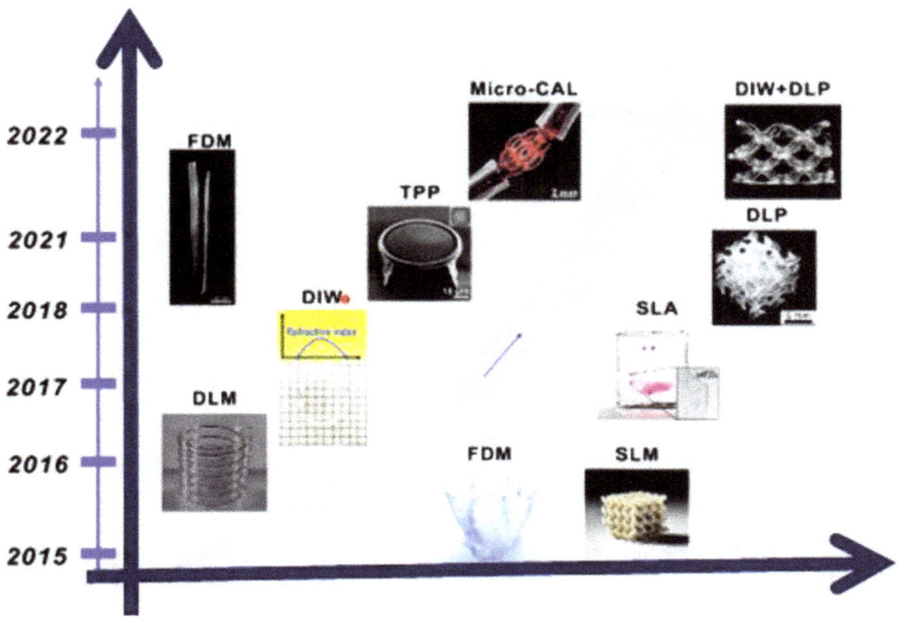

Figure 1.1. History of glass additive manufacturing. Reproduced with permission from Elsevier [14] CC BY 4.0.

Table 1.1. Classification of glass and development years. Reproduced with permission from Elsevier [14] CC BY 4.0.

3D printing technology of glass	Abbreviation	Appeared year
Fused deposition modelling	FDM	2015
Selective laser melting	SLM	2015
Directed energy deposition	DEP	2016
Direct ink writing	DIW	2017
Stereolithography	SALE	2017
Digital light projection	DLP	2018
Two-photon polymerization	TPP	2021
Fused deposition modelling	FDM	2021
Micro-computed axial lithography	MCA	2022

everyday lives, such as windows, tumblers, bottles, lights, eyeglasses, and mirrors, are referred to as common glass [16].

1.3.1.1 Soda lime glass

The most often used glass is the soda lime kind. It is used for many different things, including bakeware, bottles, jars, and containers. Because it is simple to make and can be re-melted several times, this kind of glass is sustainable. As a result, it's one of the most affordable glasses to use and ideal for recycling. Typically, SLG is made up

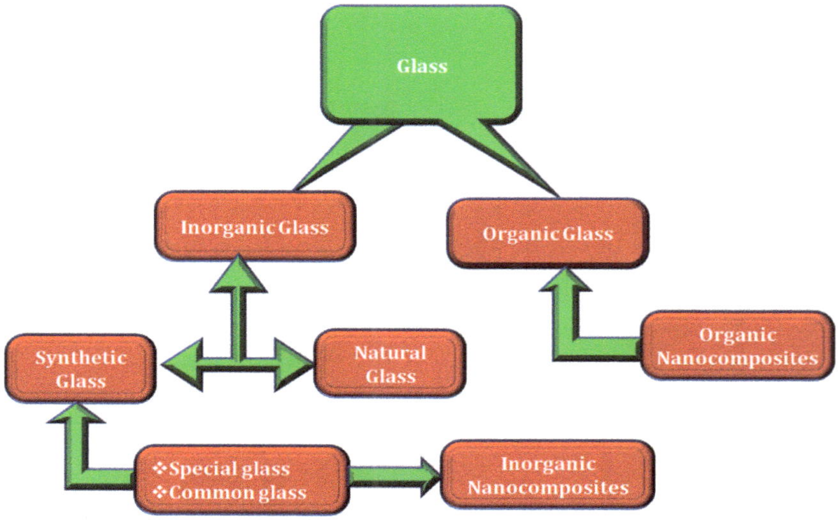

Figure 1.2. Two different types of glass materials.

of 9% lime (calcium oxide), 15% soda (sodium oxide), and 70% silicon dioxide. The remainder of its makeup might consist of many different substances [17].

1.3.1.2 Aluminosilicate glass

Compared to other forms of glass, alumina silicate glass has a higher heat tolerance. At 800 °C (1472 °F), it can withstand extreme temperatures. Lamps, high-temperature thermometers, and other items that must withstand high temperatures frequently employ it as a result of this. 5.5% silicon dioxide, 20.5% alumina, 12% magnesia, 4% boric oxide, 1% soda, and 5.5% lime (calcium oxide) are the usual ingredients of aluminosilicate glass [18].

1.3.1.3 Non-alkali or alkali-free glass

Glass for flat panel display, such as LCD or OLED (organic light-emitting diode), is either non-alkali or alkali-free. As a result, it improves resolution similarly to smartphones. Glass that isn't alkali can also be extremely thin with little distortion. Because of this, even though this kind of glass is extremely thin, it is highly resistant to harm. A variety of substances, primarily silicon dioxide, aluminum oxide, and boron trioxide, are found in non-alkali glass. It could also include trace quantities of magnesium oxide, barium oxide, strontium oxide, zinc oxide, or lime (calcium oxide) [19].

1.4 Structural formation of glass materials

The key issue in glass science is the composition of glass. The vitreous super-cooled liquid state that exists between the molten liquid state and the crystalline state distinguishes glass clearly from other forms of condensed materials. The solid, liquid, and gaseous phases found for other elements are distinct from this special characteristic of glass. The word 'glass' is typically used in science to refer to any

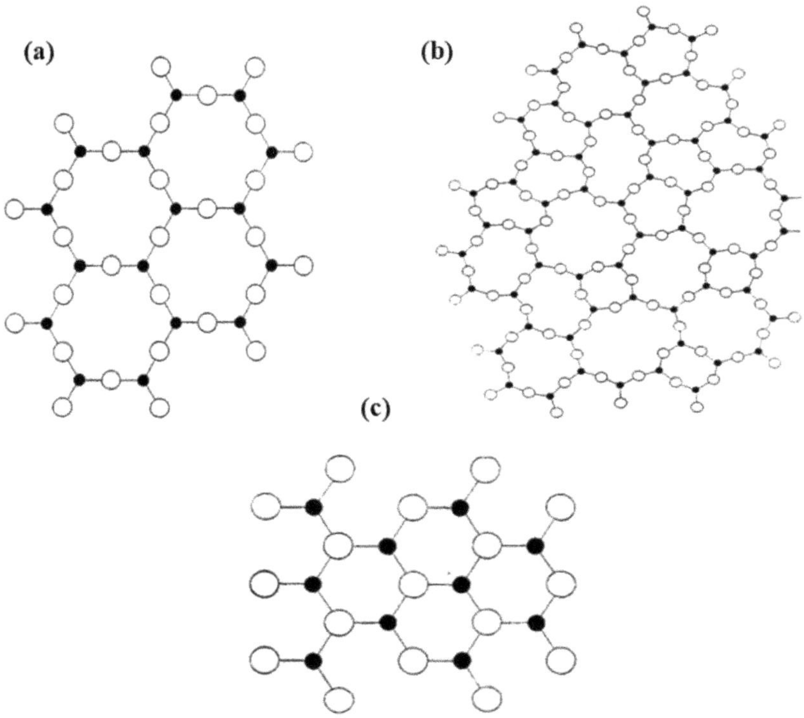

Figure 1.3. (a) A_2O_3 two-dimensional crystal's lattice is depicted, (b) while the glass compound's network is shown. (c) atom A is surrounded by three oxygen atoms and vice versa in the two-dimensional lattice of a crystal known as AO. Reprinted from [24], American Chemical Society CC BY 4.0.

solid of glass transition when heated toward the liquid state. This definition of glass goes considerably further than the conventional understanding of a see-through, silica-based solid. This broader definition states that glasses may be created from a variety of materials, including metallic alloys, ionic melts, aqueous solutions, molecular liquids, and polymers [20, 21].

All glassy states, from closely disordered configurations to liquid to closely ordered arrangements to crystalline, should be included in the so-called 'structure of glass'. It does not easily describe glass structure in terms of random network theory or microcrystalline theory due to the large variety of states. The fact that glasses are in a thermodynamically metastable condition must be considered while attempting to understand their structural form [22, 23].

Figures 1.3(a)–(c) explain the crystal structure of the random network model, glass is a three-dimensional network without symmetry or periodicity, where no structural unit repeats itself regularly. These networks consist of oxygen polyhedral for oxide glass. It is assumed that the energy content of a glass should be comparable to that of the matching crystal.

(i) No more than two glass-forming atoms should be connected to an oxygen atom.
(ii) There are four or fewer glass-forming atoms in the coordination number.

(iii) Oxygen polyhedrals share faces and corners, but not edges.
(iv) Each polyhedron should have at least three common corners in 3D networks [24].

1.4.1 Synthesis of glass materials

Glasses may be formed using a broad range of processes. The melt-quenching approach and sol–gel synthesis are two of the most widely used and simple methods for preparing various glasses, especially at the laboratory scale [25, 26].

- Chemical vapour deposition(CVD).
- Glow-discharge.
- Sputtering.
- Sol–gel synthesis [27].
- Melt-quenching.

1.5 Properties of glass materials

The following list highlights some of the essential features of glass [28]:

- it is an element created from organic materials;
- it is a piece of solid inorganic matter;
- high temperatures are not an issue for glass;
- it is completely recyclable;
- an effective electrical insulator;
- not influenced by water, air, acid, or chemical reagents;
- there is no fixed melting point.

1.5.1 Mechanical and thermal properties

The majority of glasses have high hardness values relative to other materials, but since they are fragile, they are not suitable for many technical applications. Numerous efforts have been made to increase the glass's ability to shatter. Below, a few of the mechanical and thermal characteristics of glasses are covered, along with various technical applications [29].

1.5.1.1 Viscosity

Glass viscosity is a crucial component. Most processing methods demand a viscosity between 103 and 107 dPa at a specific temperature. The working point referred to as this softening point of the glass is known to occur when its viscosity hits 107.6 dPa. Glass rapidly deforms at this stage and can no longer be utilized for sintering or glassblowing. Most glasses exhibit viscosity in a range of 1012–1013.5 dPa in the glass transformation range; the top limit, 1013 dPa, is known as the annealing point. For the strains created during the hot forming of glasses, annealing is crucial. The relaxation of stress takes longer at 30 °C–40 °C below the annealing threshold, when the viscosity is 1014.5 dPa. Compared to the annealing point (43 h) and (1 h). The strain point is where this is located. The recommended temperature for stress relief should be determined by the application and dimensional accuracy requirements [30].

1.5.2 Optical properties of glass

Refraction, absorption, and transmission of light are the three categories of optical characteristics of glasses that may be separated. Only a few other substances, including glass, can transmit light in the visible range. Therefore, the creation of optical glasses with the proper refractive index and dispersion properties is crucial to the advancement of modern research, particularly in the fields of nanotechnology, medicine, astronomy, and biology [31–33].

1.5.2.1 Absorption (UV)

A UV cut-off is effectively produced by strong absorption bands resulting from electron excitations, which makes most glasses seem opaque in the UV. There are two main kinds of these electron excitations:

(1) Intrinsic excitation, in which electrons are stimulated to inter-band transition levels in the conduction band or exaction from valence band unoccupied states.
(2) Intra-ionic transitions, the transfer of an electron (charge) from one ion's shell to another nearby ion's shell. These ions might be significant or insignificant components [34].

1.5.2.2 Photoelasticity

Glass has a random structure in all directions, it is assumed to be isotropic in a well-annealed and homogenous condition. However, the optical isotropy is disrupted when non-hydrostatic pressures are present. Different velocities are experienced by perpendicular vibrations of un-polarized light as they pass through the medium. This suggests that the glass acquires the ability to double-refract or exhibit birefringence. When seen along the x_2-axis, the optical path difference between the vibrations along the x_3- and x_1-axes is given by

$$\delta = C_\lambda \sigma_{33} t \tag{1.1}$$

where 't' is the specimen's thickness. C_λ is sometimes referred to as the stress-optic coefficient or Brewster's constant [35, 36].

1.5.2.3 Anomalous birefringence

The twofold refraction characteristic seen in phase-separated glasses even in the absence of any external or internal stress is known as anomalous birefringence. Such twofold refraction is almost often the result of a natural process, especially in glasses with distinct second-phase particles (separation via a nucleated mechanism). Form birefringence results from the existence of an anisotropic particle shape, while distribution birefringence results from an anisotropic spatial distribution. Stretched Corning code 7070 glass specimens that have undergone phase separation treatment in the past exhibit the two kinds. Discrete borate-rich particles form in a continuous silica-rich matrix within a low-alkali borosilicate glass [37].

1.6 Application and recent development of glass materials

The non-exhaustive list of items below includes products that employ glass [38].

- Packaging (food in jars, beverages in bottles, and cosmetics and drugs in flacons).
- Tableware (glasses for drinking, plates, cups, and bowls).
- Residential and commercial constructions (windows, facades, conservatories, insulation, and support structures).
- Furniture and interior design (mirrors, walls, balustrades, tables, shelves, lights).
- Transportation and automotive (windscreens, backlights, lightweight yet reinforced structural parts of vehicles, ships, airplanes, etc).
- Biotechnology, life science engineering, optical glass, and medical technology.
- Radiation protection against nuclear gamma rays and x-rays (radiology).
- Fibre optic cables are used to transmit data to phones, TVs, and computers.
- Renewable energy sources (wind turbines, solar-energy glass).
- Electricity will be provided to hybrid and electric automobiles via a photovoltaic sunroof [39].
- LED lighting fixtures for jewellery.
- Mirrors that evaluate the person in front of them for their health.
- Improved insulating characteristics and more sophisticated glass designs to release architects from limitations.
- The majority of Europe's energy demands may be met by smaller, best-performing solar and wind energy power-producing equipment with simpler integration [40].

1.6.1 Bioactiveive glass for biological performance

Many studies over the past 50 years have sought to create novel materials that can combine high performance with improved biological responsiveness. Since its discovery at the end of the 1960s, 45S5 Bioglass has undergone extensive research, opening up possibilities for bone repair and regeneration [41]. This material exhibits advantages regarding angiogenesis and antibacterial activity, can firmly attach to bone, has osteoinductive and osteoconductive abilities, and may be employed for the regulated release of therapeutic ions [25]. A novel bioactive glass using a 'melt-quenching method' was evaluated for its biological and thermal performance. The major goal was to make a unique product with a high crystallization temperature in consideration of potential thermal treatments and exceptional biological responsiveness. Through the use of a cutting-edge 3D cellular model that matched the prospective clinical use, the potential for biological activity of the novel glass. Human bone marrow mesenchymal stem cells (BM-MSCs) to investigate the bioactive glass granules performed in this 3D biological model. Human BM-MSC adhesion, colonization, and bone differentiation were promoted by the bioactive glass.

In light of this, this novel bioactive glass appears to be especially interesting for orthopedic applications, bone tissue engineering, and the field of regenerative medicine, particularly if a heat treatment is required for the creation of certain devices [39]. Composites made of bioactive glass (BAG) and poly (lactic acid) (PLA) show considerable promise for bone tissue engineering. The goal of these materials is to create a scaffold with adaptable characteristics that combines the benefits of the components of composites, such as biodegradability, bioactivity, and osteoinduction [32]. BG_Ca/Mix bioactive glass and two unique sintered bodies made of HA/BG_Ca-Mix composites are used to heal bone tissue. After 8 weeks after implantation, the 45S5 specimens developed localized cracks; this fact can be attributed to the partial crystallization of 45S5 since the remaining glass phase was likely preferentially destroyed by the physiological fluids [31].

1.6.2 Chalcogenide glass fiber technology

Chalcogenide glasses are constructed from the chalcogen elements S, Se, and Te, and stable glasses are created by the addition of additional elements including Ge, As, and Sb. There have been several potential applications for infrared fiber operate at longer wavelengths than silica fiber in the short-range area, including gas sensing, laser power transmission, remote temperature monitoring, and thermal imaging research effort has been working on creating infrared fiber that can operate in the spectral range beyond 3 μm since 1986 [42]. Sulfide, selenide, and telluride are the three systems into which chalcogenide glass is often divided. In-depth research has been done on glass-forming zones and glass characteristics. A loss of less than 0.1 $dB\,m^{-1}$ might be obtained, according to certain sulfide and selenide glasses that were pulled into the unclad fiber. However, because the fiber has a limited mechanical strength, a buffer material is required to protect it. The most common technique is Teflon (perfluorinated ethylene propylene) coating [43]. The power supply of the CO laser was carried out using As_2S_3 fiber with Teflon coating [44, 45].

1.6.2.1 Active application

In addition to the fiber's related scattering, absorption, and end-face reflection losses, the method by which light propagates through the fiber is altered. Examples of these include gratings, bright sources, fiber lasers, amplifiers, and non-linear effects.

1.6.2.2 Passive application

The fibers are used to transmit light from one place to another without interacting with the light other than to compensate for the fiber's scattering, absorption, and end-face reflection losses [46].

1.6.3 Solar cell application for thin glass

Tempered thin glass is light, incredibly flexible, and incredibly durable—perfect qualities for use in the solar industry in glass–glass modules or as cold-bent parabolic reflectors. In LiSEC, the glass–glass module encapsulation method incorporates all the benefits of tempered thin glass, using a half-century of experience in the insulating glass industry. Their airtight closure makes the modules fully UV-resistant and diffusion-tight. The use of thin glass for the front and back rail installation is made simple on the back side [47]. The method of LiSEC encapsulation is ideal for appropriate for thin-film, crystalline, and organic solar cells, and the laminating film may be selected by the consumer's needs as well [48]. Up to 6% more energy may be produced when thin glass, AR coating, and laminated film are perfectly matched. After 25 years, this represents an increase in energy production of 450 kWh for a conventional module (72 cells, 300 Wp) [49]. Longer lifespan of the strong thin glass and LiSEC's expertise in sealing methods ensure that the modules are completely UV-resistant and diffusion-tight [50]. Because of the symmetric structure of the module, the cells are protected from breaking under bending force by being located within the neutral zone of the module. Compared to traditional ones, the modules are incredibly light because of the thin glass utilized. Consequently, sub-constructions that are easier to use and more affordable may be employed [51, 52]. Figure 1.4 explains that SiNW-based solar cells might have power conversion efficiency as high as 20%. Typically, when the term SiNWs is used, structures are produced using the growth process of vapour, liquid, and solid (VLS). Wet electroless chemical etching of 2.7 m multicrystalline p^+nn^+ treated silicon layers using silver nitrate and hydrofluoric acid was used to create the nanowire structure for silicon nanowire-based solar cells over glass substrates. It has been measured to have a large broadband optical absorption (>90% across 500 nm) and low reflectance (<10%, at 300–800 nm). At an optimal power conversion efficiency of 4.4%, the maximum open-circuit voltage (V_{oc}) and short-circuit current densities (J_{sc}) for AM1.5 lighting were 450 mV and 40 mA cm^{-2}, respectively [53].

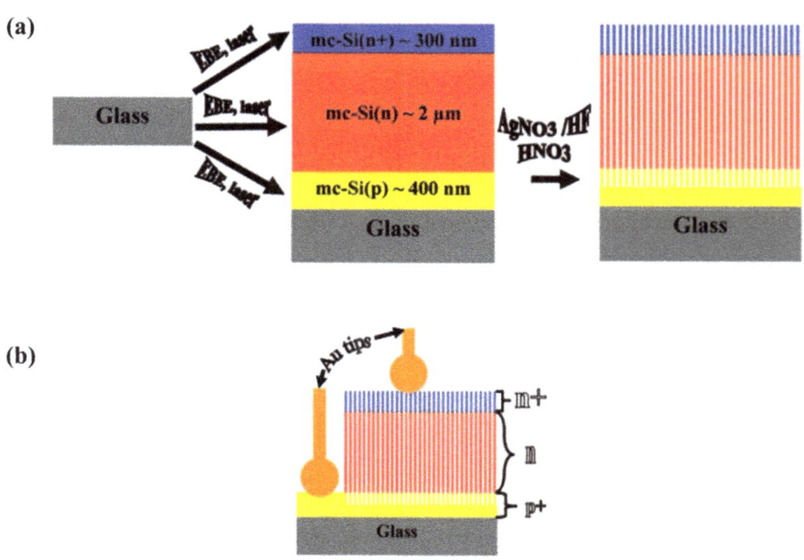

Figure 1.4. (a) Diagrammatic cross-sectional image of the SiNWs following wet chemical etching (right) and the mc-Si p–n junction in a layers stack over a glass substrate (left). (b) Diagrammatic depiction of SiNW-based p–n junction I–V curve measurements. Reprinted from [53], American Chemical Society. CC BY 4.0.

1.6.3.1 Porous silicon glass

Monolithic solar modules with superior thin-film crystalline silicon layers on glass substrates have the potential to significantly reduce the cost of photovoltaic power. The cost of the silicon wafers themselves presently accounts for around one-third of the price of wafer-based silicon solar panels. There are many benefits to using thin crystalline silicon films on cheap substrates as the active material rather than silicon wafers:

- The first is that a significant reduction in silicon usage should result in significant cost savings.
- Second, compared to wafer-based cells, high cell efficiencies may be achieved using silicon material that has significantly poorer crystalline quality than the utilization of extremely thin silicon layers.

Three different strategies that are currently being explored to address the challenge of forming high-quality thin-film crystalline silicon layers on glass substrates could significantly reduce the price of photovoltaic electricity. Monolithic solar modules made from these layers could significantly lower the price of photovoltaic electricity. A wafering method called SLiM-Cut produces layers that are 50 microns thick with the least amount of material loss possible. The process of lifting off ultra-thin monocrystalline films created by silicon's cylindrical macrospore arrays reorganizing during annealing is known as the epifree method. Thus far, extremely basic cells with an efficiency of up to 4.1% and films with a thickness of around 1 mm have been accomplished [54]. A review of the fundamental issues with conventional thin films—namely, toxicity, resource availability, complexity/cost trade-off, and stability—led to the inception of the crystalline silicon on glass (CSG) project in the late 1980s [50]. The establishment of Pacific Solar in 1995 gave the program the crucial impetus it needed to get the technology ready for the market. An additional justification for the program was the advancements in silicon technology throughout the 1980s with the exceptional durability probably resulting from the almost perfect materials system (silicon, its oxide, and its nitride) provided by the CSG approach. The discovery that light-trapping inside the film might theoretically enable great efficiency from silicon films as thin as a micron or less was particularly significant in the former field [55].

1.6.3.2 Crystalline silicon on glass technology

Among thin-film technologies, the CSG solar cell technology is among the most similar to the most successful crystalline silicon (c-Si) wafer-based photovoltaics [56]. It seeks to combine the advantages of large-area monolithic cell-module integration, inline-capable processing typical of thin films, cheap material consumption, and technology of c-Si wafer cells [57]. Furthermore, because smaller devices are projected to have a higher cell voltage and can still gather minority carriers with a shorter diffusion length, c-Si thin-film cells could benefit from less strict criteria for the material's electrical quality [58]. Commercial CSG solar modules were manufactured by CSG Solar AG, with an efficiency of 6%–7% and a record-breaking 10.4%. Solid-state crystallization (SPC) of 1.5–3.5 mm thick precursor diodes made by PECVD or beam evaporation, followed by thermal annealing, hydrogen passivation, and metallization, is the process used to create CSG cells [59]. On textured borosilicate glass, a PECVD mini-module exhibited a maximum efficiency of 10.4%. On the planar glass, the best-performing beam-evaporated cells achieved an efficiency of 8.6%. Additionally, CSG cells with 8.1% and 7.1% efficiency on PECVD and beam material were generated on inexpensive SLG [60].

1.6.3.3 Soda lime glasses

SLGs have a typical composition (wt%) of 70 SiO_2, 10 CaO, 15 Na_2O [61], and small amounts of other oxides. These glasses are employed for high-volume products such as windows, bottles, and jars [62]. As a result, the cross-linking is decreased when soda (NaO_2) is introduced to silica glass because each Na^+ ion attaches itself to an oxygen ion of a tetrahedron [63]. Thus, the addition of soda has the effect of substituting (non-directional) ionic bonds with lower energy for some of the covalent bonds that hold the tetrahedral together. As a result, the melt becomes less viscous, making soda glass workable at around 700 °C, while pure silica softens at roughly 1200 °C. Conversely, the glass's ability to withstand high temperatures is diminished by the alloying process, which is why silica glass is only appropriate for use in high-temperature applications like quartz halogen lamp enclosures [64, 65].

More recently, technical developments have led to the creation of thin-film glasses for substrates that are stronger, more diffusive, and more thermally and mechanically stable than SLG that is sold commercially [66]. The current glasses have qualities that make them appropriate for usage as a thin-film solar cell substrate. As absorber layers in thin-film solar cells, thin films of ternary semiconductors like $CuInGaSe_2$ (CIGS) and, more recently, Cu_2ZnSnS_4 (CZTS), have attracted a lot of scientific interest [67]. It's interesting to note that in addition to the characteristics of the films resulting from differences in the parameters of the film deposition procedures, the substrate has been determined to be crucial to the solar cells' performance [57, 68]. The efficiency improvement in terms of raising carrier concentration and, consequently, electrical conductivity in the CIGSe layer is caused by Na from SLG. Subsequently, other methods for externally integrating Na into the CIGSe layer were investigated, as the amount of Na integrated from SLG could not be regulated by adding alkali metals to solar cells based on chalcogenides [69]. The 'post-deposition treatment' (PDT), which involves depositing around 30 nm of NaF over as-grown CIGSe absorbers, was one of the final techniques to be developed [70]. PDT produced solar cells that were on a par with those made using SLG substrates. By contrasting their device performances, the same scientists examined the structural impacts of Na from SLG and PDT on CIGSe produced with different substrate temperatures [69, 71].

1.6.4 Bone tissue engineering for ceramic glass

Large-scale bioactive glass-ceramic scaffolds using a polyurethane sponge template were created for tissue engineering. The beginning glass (CEL2) was created using a traditional melting-quenching method and is a member of the SiO_2–P_2O_5–CaO–MgO–Na_2O–K_2O system. A slurry consisting of CEL2 powder, polyvinyl alcohol, and water has been created to cover the polymeric template by impregnation. A glass-ceramic duplicate of the template was created by sintering the glass powders and removing the sponge using an ideal heat treatment. The obtained devices are good candidates as scaffolds for bone tissue engineering in terms of pore-size distribution, pore interconnection, surface roughness, and both bioactivity and biocompatibility, according to morphological observations, image analyses, mechanical tests, and *in vitro* tests [72, 73]. Regenerating and repairing massive bone abnormalities caused by trauma or illness is still a major therapeutic problem. Bioactive glass presents attractive properties as a scaffold material for bone tissue creation, with the use of glass scaffolds for load bearing. Bone flaws are frequently restricted due to their poor fracture toughness and mechanical strength. Numerous techniques have been employed to create scaffolds with compressive strengths that are like those of trabecular and cortical bones. The low fracture toughness (low resistance to fracture) and inadequate mechanical dependability of bioactive glass scaffolds are currently major drawbacks, which have not garnered much attention up to this point. The creation of robust and durable bioactive glass scaffolds and their assessment in animal models with unloaded and load-bearing bone defects should be the focus of future studies [74–76].

1.6.4.1 Types of bioactive glasses

Glasses composed of the same building blocks with distinct patterns, operating as amorphous solids without long-range organization, are optically transparent (often), and are brittle. Glasses are more like super-stiff liquids with more randomly ordered atoms than liquids and display how glass changes across a range of temperatures over time. The following categories apply to BGs based on features of their foundation [77, 78].

1.6.4.1.1 Silicate-bioactive glasses

To achieve the requisite bioactive characteristics, bioactive glasses consisting of borosilicate and borate have been treated with varying proportions of B_2O_3 [79]. In comparison to their silica counterparts, borate/borosilicate glasses (BSGs) react more quickly and completely to apatite [80]. The sintering behavior of BSGs is also more controlled. The conversion mechanism via the borate-rich layer is like that of silicate-bioactive glass. The microstructure of bioactive glasses made of borosilicate and borate (like 13-93B2) is almost the same as that of trabecular bone [81, 82].

1.6.4.1.2 Phosphate-bioactive glasses

The phosphate (PO_4) 32 tetrahedron structural unit of phosphate-bioactive glasses is asymmetric, and P_2O_5 functions as a network-forming oxide [83]. Low chemical durability is caused by this asymmetry, which shows that P–O–P bonds may be easily hydrated [84]. Phosphate-bioactive glasses can have their dissolving rate controlled by adding the right metal oxides (such as TiO_2, CuO, NiO, MnO, and Fe_2O_3) to the glass composition [85]. As a potential bone material, phosphate-bioactive glasses (such as Na_2O–CaO–SrO–P_2O_5) have demonstrated good physical and structural characterization [86–88].

1.6.4.1.3 Doped-bioactive glasses

Doped-bioactive glasses (DBGs) are glass compositions that have been changed by the addition of dopants or other additives to make them bioactive, bioresorbable, and/or biodegradable [89]. Doping elements including Al, Cu, Zn, Fe, Cr, In, Ba, Sr, and *Y* have been widely used to functionalize BGs [90, 91]. Ag-doped DBG compositions demonstrated effective antibacterial characteristics while retaining their bioactive nature [92], moreover, in addition to SiO_2, B_2O_3, and P_2O_5, additional oxides including Na_2O, K_2O, CaO, and MgO help to balance the pH of the surrounding environment [93]. CuO, ZnO, AgO, and TiO_2 also help to release the right ions that can impart an antibacterial impact on the biomaterial [81, 94]. While Zn and Mg are generally recognized to have a stimulatory effect, Al_2O_3 helps enhance the mechanical characteristics of the BGs [95].

1.6.4.1.4 Metallic-bioactive glasses

Bulk metallic glasses, or BMGs, are biodegradable without the development of hydrogel and have special qualities including higher mechanical strength, high fracture toughness, high elastic strain limit, and reduced Young's modulus. BMGs exhibit better qualities than traditional biomaterials. Perhaps the primary drawback of BMGs is the addition of nickel (Ni), which can trigger an allergic reaction and could act as a carcinogenic component [90].

1.7 Conclusion and future scope

Finally, this chapter concludes that glass materials have high transparency, minimum surface roughness, high accuracy, and cheap energy for the fabrication of small-scale structures making this technology ideal for usage in the glass sector. This technical advancement also satisfies the various new requirements for glass. It is a resourceful, entirely recyclable material with significant environmental benefits that has been used from the Stone Age to now. The historical background and future scope of glass materials are discussed.

1.8 Some R&D initiatives for the development of glass in the future scope

Many more fundamental studies are needed; particularly on the framework of the glass surface, the chemical processes occurring there, and the way these reactive areas interact with molecules, to further improve glass qualities. We require hard science and a breakthrough in our knowledge of the composition of glass. Although it demands significant capital, a successful research project might lead to everlasting possibilities.

1. Building-specific anti-reflection characteristics for the production of renewable energy.
2. Strength: If the glass is 50 times stronger than it is now, new goods and business prospects may materialize on the market, such as light and thin container glass, lighter flat glass, and fiberglass for composites. Some applications already increase the strength of glass by two to six times.

3. Glass that has been functionally integrated is consequently a perfect substrate for OLED lights, touch panels, audiovisual displays, etc.
4. Additionally, there is thinner glass, scratch thinner, scratch-resistant glass-resistant, audio glass, and flexible [40].

References

[1] Sen S 2022 The world of pilgrims *Ganges* (Yale University Press) ch 1 pp 14–44

[2] Choi Y and Kim Y 2021 Applications of the smart helmet in applied sciences: a systematic review *Appl. Sci.* **11**

[3] Azouz A B, Vázquez M and Brabazon D 2014 Developments of laser fabrication methods for lab-on-a-chip microfluidic multisensing devices *Compr. Mater. Process.* **13** 447–58

[4] Berenjian A and Whittleston G 2017 History and manufacturing of glass *Am. J. Mat. Sci.* **2017** 18–24

[5] Haldimann M, Luible A and Overend M 2008 Structural use of glass *Structural Engineering Documents (No. 10)* International Association for Bridge and Structural Engineering (IABSE), Zurich, Switzerland

[6] Overend M, Jin Q and Watson J 2011 The selection and performance of adhesives for a steelglass connection *Int. J. Adhes. Adhes.* **31** 587–97

[7] Ahmadi M, Wu T and Hu B 2017 A review on organic–inorganic halide perovskite photodetectors: device engineering and fundamental physics *Adv. Mater.* **29** 1–24

[8] Fang Y, Dong Q, Shao Y, Yuan Y and Huang J 2015 Highly narrowband perovskite single-crystal photodetectors enabled by surface-charge recombination *Nat. Photonics* **9** 679–86

[9] Grätzel M 2014 The light and shade of perovskite solar cells *Nat. Mater.* **13** 838–42

[10] Kurkjian C R and Prindle W R 1998 Perspectives on the history of glass composition *J. Am. Ceram. Soc.* **81** 795–813

[11] Axinte E 2011 Glasses as engineering materials: a review *Mater. Des.* **32** 1717–32

[12] Hutter K, Ursescu A and van de Ven A A F 2006 General introduction *Lect. Notes Phys.* **710** 1–6

[13] Montazerian M and Dutra Zanotto E 2016 History and trends of bioactive glass-ceramics *J. Biomed. Mater. Res.—Part* A **104** 1231–49

[14] Xin C, Li Z, Hao L and Li Y 2023 A comprehensive review on additive manufacturing of glass: recent progress and future outlook *Mater. Des.* **227**

[15] Scholes S R 1945 Abrasion of glass as related to composition *J. Am. Ceram. Soc.* **28** 133–6

[16] Karmakar B 2016 *Fundamentals of Glass and Glass Nanocomposites* (Amsterdam: Elsevier)

[17] De Juan Ares J, Vigil-Escalera Guirado A, Cáceres Gutiérrez Y and Schibille N 2019 Changes in the supply of eastern Mediterranean glasses to Visigothic Spain *J. Archaeol. Sci.* **107** 23–31

[18] Huang S, Wang W, Jiang H, Zhao H and Ma Y 2022 Network structure and properties of lithium aluminosilicate glass *Materials (Basel)* **15** 4555

[19] Hasanuzzaman M, Rafferty A, Sajjia M and Olabi A-G 2016 Properties of glass materials *Ref. Modul. Mater. Sci. Mater. Eng.* 1–12

[20] Weller B, Nicklisch F, Prautzsch V, Döbbel F and Rücker S 2010 All glass enclosure with transparently bonded glass frames *Challenging Glas. 2—Conf. Archit. Struct. Appl. Glas. CGC 2010* pp 207–16

[21] Jiang Z and Zhang Q 2014 Progress in materials science the structure of glass: a phase equilibrium diagram approach *J. Prog. Mater. Sci.* **61** 144–215

[22] Brow R K and Tallant D R 1997 Structural design of sealing glasses *J. Non-Crystal. Solids* **222** 396–406

[23] Varshneya A K and Mauro J C 2019 Glass compositions and structures *Fundamentals of Inorganic Glasses* **vol 2** 3rd edn (Elsevier) ch 5

[24] Zachariasen W H 1932 The atomic arrangement in glass *J. Am. Chem. Soc.* **54** 3841–51

[25] Jones J R 2015 Reprint of review of bioactive glass: from hench to hybrids *Acta Biomater.* **23** S53–82

[26] Kothiyal G P, Ananthanarayanan A and Dey G K 2011 *Glass and Glass-Ceramics* (Amsterdam: Elsevier)

[27] Zhang X and Cresswell M 2016 Materials for inorganic controlled release technology *Inorganic Controlled Release Technology* (Elsevier) pp 1–16

[28] Hasanuzzaman M *et al* 2016 *Properties of Glass Materials* (Reference Module in Materials Science and Materials Engineering) (Elsevier) pp 1–12

[29] Hoffmann H and Universität T 1932 Optical glasses *Nature* **129** 584

[30] Zachariasen W H 1932 The atomic arrangement in glass *J. Am. Chem. Soc.* **54** 3841–51

[31] Bellucci D, Anesi A, Salvatori R, Chiarini L and Cannillo V 2017 A comparative *in vivo* evaluation of bioactive glasses and bioactive glass-based composites for bone tissue repair *Mater. Sci. Eng.* C **79** 286–95

[32] Houaoui A, Lyyra I, Agniel R, Pauthe E, Massera J and Boissière M 2020 Dissolution, bioactivity and osteogenic properties of composites based on polymer and silicate or borosilicate bioactive glass *Mater. Sci. Eng.* C **107** 110340

[33] Ahani A and Ahani E 2023 An overview for materials and design methods used for enhancement of laminated glass *Hybrid Adv* **3** 100063

[34] Rajaramakrishna R and Kaewkhao J 2019 Glass material and their advanced applications *KnE Soc. Sci.* **2019** 796–807

[35] Kittel C and Holcomb D F 1967 Introduction to solid state physics *Am. J. Phys.* **35** 547–8

[36] Varshneya A K and Mauro J C 2019 Fundamentals of inorganic glass making *Fundamentals of Inorganic Glasses* (Elsevier) ch 22

[37] When V 1986 Optical properties *Developments in Agricultural Engineering* **vol 8** (Elsevier) ch 6

[38] Jóźwik A 2022 Application of glass structures in architectural shaping of all-glass pavilions, extensions, and links *Buildings* **12** 1254

[39] Bellucci D, Veronesi E, Dominici M and Cannillo V 2020 A new bioactive glass with extremely high crystallization temperature and outstanding biological performance *Mater. Sci. Eng.* C **110** 110699

[40] Glass Alliance Europe 2012 Glass sustainability and the environment—Glass Alliance Europe [Online] https://glassallianceeurope.eu/en/environment

[41] Stefanic M *et al* 2018 The influence of strontium release rate from bioactive phosphate glasses on osteogenic differentiation of human mesenchymal stem cells *J. Eur. Ceram. Soc.* **38** 887–97

[42] Nishii J, Morimoto S, Inagawa I, Iizuka R, Yamashita T and Yamagishi T 1992 Recent advances and trends in chalcogenide glass fiber technology: a review *J. Non. Cryst. Solids* **140** 199–208

[43] Vigreux C *et al* 2014 Wide-range transmitting chalcogenide films and development of micro-components for infrared integrated optics applications *Opt. Mater. Express* **4** 1617

[44] Richet P, Conradt R, Takada A and Dyon J (ed) 2021 *Encyclopedia of Glass Science, Technology, History, and Culture* (Wiley)

[45] Bordas S, Clavaguera-Mora M T and Clavaguera N 1990 Glass formation and crystallization kinetics of some GeSbSe glasses *J. Non. Cryst. Solids* **119** 232–7

[46] Sanghera J S, Shaw L B and Aggarwal I D 2002 Applications of chalcogenide glass optical fibers *C. R. Chim* **5** 873–83

[47] Máčalová K, Václavík V, Dvorský T, Figmig R, Charvát J and Lupták M 2021 The use of glass from photovoltaic panels at the end of their life cycle in cement composites *Materials (Basel)* **14** 6655

[48] O'Regan B and Grätzel M 1991 A low-cost, high-efficiency solar cell based on dye-sensitized colloidal TiO2 films *Nature* **353** 737–40

[49] Mohammad Bagher A 2015 Types of solar cells and application *Am. J. Opt. Photonics* **3** 94

[50] Widenborg P I and Aberle A G 2007 Polycrystalline silicon thin-film solar cells on AIT-textured glass superstrates *Adv. OptoElectron.* **2007** 24584

[51] Karasu B, Bereket O, Biryan E and Sanoğlu D 2017 The latest developments in glass science and technology *El-Cezeri Fen ve Mühendislik Derg* **4** 209–33

[52] Elbashar Y H, Rayan D A, Saleh H E R, Rasoul A and Roshdy A 2021 Glass technology and its application in solar cells *Nonlin. Opt. Quantum Opt. Concepts Mod. Opt.* **53** 177

[53] Sivakov V *et al* 2009 Silicon nanowire-based solar cells on glass: synthesis, optical properties, and cell parameters *Nano Lett.* **9** 1549–54

[54] Gordon I *et al* 2011 Three novel ways of making thin-film crystalline-silicon layers on glass for solar cell applications *Sol. Energy Mater. Sol. Cells* **95** 2–7

[55] Green M A *et al* 2004 Crystalline silicon on glass (CSG) thin-film solar cell modules *Sol. Energy* **77** 857–63

[56] Green M A 2009 Polycrystalline silicon on glass for thin-film solar cells *Appl. Phys. A Mater. Sci. Process.* **96** 153–9

[57] Barkhouse D A R, Gunawan O, Gokmen T, Todorov T K and Mitzi D B 2015 Yield predictions for photovoltaic power plants: empirical validation, recent advances and remaining uncertainties *Prog. Photovolt. Res. Appl.* **20** 6–11

[58] Young D L *et al* 2010 Toward film-silicon solar cells on display glass *Conf. Rec. IEEE Photovolt. Spec. Conf.* 626–30

[59] Jin G, Widenborg P I, Campbell P and Varlamov S 2010 Lambertian matched absorption enhancement in PECVD poly-Si thin film on aluminum-induced textured glass superstrates for solar cell applications *Prog. Photovoltaics Res. Appl.* **18** 582–9

[60] Varlamov S *et al* 2013 Polycrystalline silicon on glass thin-film solar cells: a transition from solid-phase to liquid-phase crystallized silicon *Sol. Energy Mater. Sol. Cells* **119** 246–55

[61] Saha R and Nix W D 2002 Effects of the substrate on the determination of thin film mechanical properties by nanoindentation *Acta Mater.* **50** 23–38

[62] Wei S H, Zhang S B and Zunger A 1999 Effects of Na on the electrical and structural properties of CuInSe2 *J. Appl. Phys.* **85** 7214–8

[63] Bansal N, Pandey K, Singh K and Mohanty B C 2019 Growth control of molybdenum thin films with simultaneously improved adhesion and conductivity via sputtering for thin film solar cell application *Vacuum* **161** 347–52

[64] Martin J W 2006 Glasses and ceramics *Mater. Eng.* 133–58

[65] Couzinie-Devy F, Cadel E, Barreau N, Arzel L and Pareige P 2015 Na distribution in Cu(In, Ga)Se2 thin films: investigation by atom probe tomography *Scr. Mater.* **104** 83–6

[66] Bansal N, Chandra Mohanty B and Singh K 2020 Designing composition tuned glasses with enhanced properties for use as substrate in Cu2ZnSnS4 based thin film solar cells *J. Alloys Compd.* **819** 152984
[67] Margulis L, Hodes G, Jakubowicz A and Cohen D 1989 Aggregate structure in CuBSe2/Mo films (B=In, Ga): its relation to their electrical activity *J. Appl. Phys.* **66** 3554–9
[68] Li J V, Kuciauskas D, Young M R and Repins I L 2013 Effects of sodium incorporation in co-evaporated Cu2ZnSnSe4 thin-film solar cells *Appl. Phys. Lett.* **102** 1–5
[69] Majumdar I, Parvan V, Greiner D, Schlatmann R and Lauermann I 2020 Effect of Na from soda-lime glass substrate and as post-deposition on Cu(In, Ga)Se2 absorbers: a photoelectron spectroscopy study in ultra-high vacuum *Appl. Surf. Sci.* **514** 145941
[70] Bubli I *et al* 2020 Enhancement of solar cell efficiency via luminescent downshifting by an optimized coverglass *Ceram. Int.* **46** 2110–5
[71] Salomé P M P, Rodriguez-Alvarez H and Sadewasser S 2015 Incorporation of alkali metals in chalcogenide solar cells *Sol. Energy Mater. Sol. Cells* **143** 9–20
[72] Khater G A, Safwat E M, Kang J, Yue Y and Khater A G A 2020 Some types of glass-ceramic materials and their applications *Int. J. Res. Stud. Sci. Eng. Technol.* **7** 2349–476
[73] Huang K *et al* 2019 Halloysite nanotube based scaffold for enhanced bone regeneration *ACS Biomater. Sci. Eng.* **5** 4037–47
[74] Rahaman M N *et al* 2011 Bioactive glass in tissue engineering *Acta Biomater.* **7** 2355–73
[75] López-Noriega A, Arcos D, Izquierdo-Barba I, Sakamoto Y, Terasaki O and Vallet-Regí M 2006 Ordered mesoporous bioactive glasses for bone tissue regeneration *Chem. Mater.* **18** 3137–44
[76] Hoppe A *et al* 2014 Cobalt-releasing 1393 bioactive glass-derived scaffolds for bone tissue engineering applications *ACS Appl. Mater. Interfaces* **6** 2865–77
[77] Kumar A and SooHan S 2019 *Bioactive Glass-based Composites in Bone Tissue Engineering: Synthesis, Processing, and Cellular Responses* (Amsterdam: Elsevier)
[78] Qin C *et al* 2016 Novel bioactive Fe-based metallic glasses with excellent apatite-forming ability *Mater. Sci. Eng.* C **69** 513–21
[79] Yao A, Wang D, Huang W, Fu Q, Rahaman M N and Day D E 2007 *In vitro* bioactive characteristics of borate-based glasses with controllable degradation behavior *J. Am. Ceram. Soc.* **90** 303–6
[80] Huang W, Day D E, Kittiratanapiboon K and Rahaman M N 2006 Kinetics and mechanisms of the conversion of silicate (45S5), borate, and borosilicate glasses to hydroxyapatite in dilute phosphate solutions *J. Mater. Sci. Mater. Med.* **17** 583–96
[81] Hulsen D J W, Geurts J, van Gestel N A P, van Rietbergen B and Arts J J 2016 Mechanical behavior of bioactive glass granules and morselized cancellous bone allograft in load bearing defects *J. Biomech.* **49** 1121–7
[82] Sharifi E *et al* 2016 Preparation of a biomimetic composite scaffold from gelatin/collagen and bioactive glass fibers for bone tissue engineering *Mater. Sci. Eng.* C **59** 533–41
[83] Abou Neel E A *et al* 2005 Effect of iron on the surface, degradation and ion release properties of phosphate-based glass fibers *Acta Biomater.* **1** 553–63
[84] Abou Neel E A, Mizoguchi T, Ito M, Bitar M, Salih V and Knowles J C 2007 *In vitro* bioactivity and gene expression by cells cultured on titanium dioxide doped phosphate-based glasses *Biomaterials* **28** 2967–77

[85] Abou Neel E A, Chrzanowski W and Knowles J C 2014 Biological performance of titania containing phosphate-based glasses for bone tissue engineering applications *Mater. Sci. Eng.* C **35** 307–13

[86] Abou Neel E A *et al* 2009 Structure and properties of strontium-doped phosphate-based glasses *J. R. Soc. Interface* **6** 435–46

[87] Nicolini V *et al* 2015 Evidence of catalase mimetic activity in Ce3+/Ce4+ doped bioactive glasses *J. Phys. Chem.* B **119** 4009–19

[88] Pourhaghgouy M, Zamanian A, Shahrezaee M and Masouleh M P 2016 Physicochemical properties and bioactivity of freeze-cast chitosan nanocomposite scaffolds reinforced with bioactive glass *Mater. Sci. Eng.* C **58** 180–6

[89] Horton J A and Parsell D E 2003 Biomedical potential of a zirconium-based bulk metallic glass *Mater. Res. Soc. Symp.—Proc.* **754** 179–84

[90] Wang W H 2009 Bulk metallic glasses with functional physical properties *Adv. Mater.* **21** 4524–44

[91] Ojansivu M *et al* 2015 Bioactive glass ions as strong enhancers of osteogenic differentiation in human adipose stem cells *Acta Biomater.* **21** 190–203

[92] Quintero F *et al* 2009 Laser spinning of bioactive glass nanofibers *Adv. Funct. Mater.* **19** 3084–90

[93] Wang X, Li X, Ito A and Sogo Y 2011 Synthesis and characterization of hierarchically macroporous and mesoporous $CaO–MO–SiO_2–P_2O_5$ (M = Mg, Zn, Sr) bioactive glass scaffolds *Acta Biomater.* **7** 3638–44

[94] Stanley H R *et al* 1977 Implantation of natural tooth form bioglasses in baboons *Implantologist* **1** 34–47

[95] Navarro M *et al* 2004 New macroporous calcium phosphate glass ceramic for guided bone regeneration *Biomaterials* **25** 4233–41

Chapter 2

Physio-chemical properties of glass and glass ceramics-based materials

Saravanan Alamelu, Kamalesh Balakumar Venkatesan, Kalist Shagirtha, Manoj Kumar Srinivasan and Pugalendhi Pachaiappan

In recent years, the utilization of glass and glass ceramic-based materials have increased with distinct properties appropriate for various applications. This chapter provides a thorough exploration of glasses and glass ceramics, discussing their compositions and properties. It highlights the fundamental mechanisms of glass crystallization, as well as the controlled crystallization process leading to the development of glass ceramics. The challenges and achievements in creating glass ceramics with enhanced toughness and transparency are also examined. Additionally, the chapter emphasizes the various functionalities of glasses and glass ceramics, such as optical, thermal, mechanical, electrical, magnetic, biological, and chemical properties, along with their environmental compatibility and recyclability. In summary, this thorough chapter illuminates the intricate and dynamic physio-chemical characteristics of glasses and glass ceramics, enhancing our comprehension and progress in utilizing these materials across various practical applications.

2.1 Introduction

Glass and glass ceramics, which are distinct materials, have crucial applications across a range of industries, including consumer products, manufacturing, healthcare, and construction [1]. Their unique features and diverse applications have significantly impacted the design, production, and utilization of a wide array of items [2]. The evolution of these materials, marked by innovation, inventiveness, and scientific discoveries, spans from the early days of glassmaking to contemporary advancements in glass technology [3].

The history of glass dates back to 3500 BC in Mesopotamia and ancient Egypt, as indicated by the earliest records of glass manufacture [4]. Early glassmakers discovered that heating certain materials, like silica, yielded a special translucent

doi:10.1088/978-0-7503-5904-7ch2

material that could be molded into various shapes [5]. The craft of glassmaking progressed over the ages, with the Roman Empire playing a pivotal role in advancing glass technology. A significant milestone occurred in the first century BC with the development of glassblowing techniques, enabling the mass production of decorative items, containers, and glass vessels [6]. In contrast, glass ceramics are a relatively recent innovation, emerging in the mid-20th century. Discovered in 1953 by Dr S Donald Stookey, an influential American chemist at Corning Glass Works, glass ceramics combine the strengths of ceramics and glass, resulting in a material with enhanced strength, thermal resistance, and aesthetic appeal. Since then, glass ceramics have become indispensable in applications such as technical ceramics, cookware, and dental materials [7].

Glass, an amorphous substance with a non-crystalline structure primarily composed of silica and additional ingredients, lacks the long-range order characteristic of crystalline solids, giving it a transparent or translucent appearance [8]. Depending on its composition and production technique, glass can exhibit various characteristics, including excellent electrical insulation, low heat conductivity, and high transparency. These qualities make it ideal for use in windows, containers, lenses, and optical fibers [9]. In contrast, glass ceramics combine ceramic and glass qualities, featuring a crystalline structure within a glassy matrix. Controlled crystallization of specialized compounds within the glass matrix produces fine-grained crystals with remarkable chemical resistance, mechanical strength, and heat shock tolerance. Glass ceramics can also be designed for diverse colors, textures, and surface finishes, making them suitable for utilitarian and decorative purposes in both industrial and domestic settings [10].

The adaptability of glass and glass ceramics has led to their extensive use in various industries and everyday products. Glass finds applications in construction for structural elements, architectural glazing, and facades, contributing to the creation of contemporary, aesthetically pleasing structures with abundant natural light [11]. Glass ceramics, on the other hand, are utilized in cooktops and cookware due to their exceptional resistance to heat and scratches, making them ideal for food preparation and cooking [12]. Additionally, bioactive glass is valuable in the medical and dental fields as it can form a bond with live tissues and promote bone regeneration, advancing the development of dental implants and bone grafts [13]. In the electronics sector, glass is extensively used in the production of semiconductors, optical fibers, and display panels, facilitating the creation of high-resolution screens and communication networks [14]. Glass ceramics, with their outstanding mechanical and thermal properties, are employed in aerospace components, including thermal protection systems for supersonic aircraft and heat-resistant tiles for spacecraft re-entry vehicles [15]. Figure 2.1 provides a visual representation of some of the diverse fields where glass ceramics find application.

This chapter examines the fundamental concepts and features of glasses and glass ceramics, emphasizing their pivotal role in diverse fields such as biomedical engineering, electronics, and construction. It underscores the intricate relationship between their composition and temperature history, a crucial factor for tailoring their properties to specific applications. The discussion covers the manufacturing of

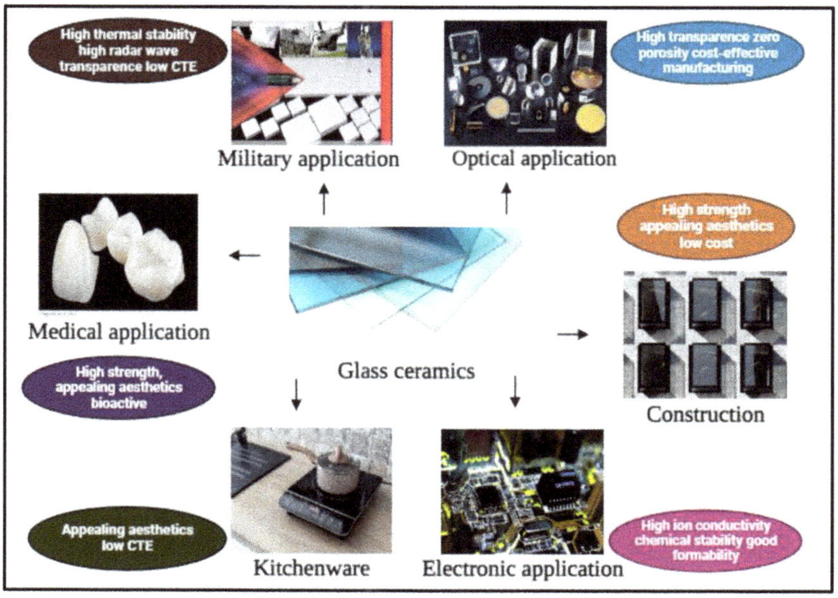

Figure 2.1. Applications of glass ceramics in a wide range of fields.

glasses from different compounds, their evolution into glass ceramics, and the associated processes of controlled crystallization. Additionally, the chapter explores the extensive range of attributes exhibited by glasses and glass ceramics, encompassing mechanical, electrical, magnetic, optical, biological, and chemical properties, underscoring their advantages across various domains. It also sheds light on ongoing research endeavors, highlighting the persistent challenges in achieving high toughness while preserving transparency.

2.2 Glasses

A widely accepted yet fundamentally accurate concept posits that glass is essentially a liquid devoid of flow: any liquid with slow crystallization kinetics will experience structural arrest, resulting in a glassy appearance [16]. However, the glassy state is not uncommon. The specific glassy 'state' achieved is determined by the temperature history of the melt. Certain binary substances—whether oxide or non-oxide—have the potential to serve as effective glass formers. Nevertheless, additional components are often introduced to enhance their suitability for various applications. Consequently, technical glasses incorporate conditional network formers (e.g., Al_2O_3, ZnO), modifiers (e.g., Li_2O, Na_2O), and occasionally refining agents (e.g., As_2O_3), alongside the primary glass former (e.g., P_2O_5, SiO_2, B_2O_3). These components aid in refining the melt and minimizing bubbles in the final glass product. Extensive literature addresses the roles of different oxides and additives [17].

An adept glass artisan operates within a context where the disordered molecular arrangements inherently possess low energy, and distinct modes are separated by high-energy barriers, without delving into the complexities of glassy compounds.

When rapid cooling occurs to prevent nucleation, the probability of crystal formation during the vitrification process is minimal, as the crystal state consistently maintains lower energy. Within a specific timeframe known as the nucleation time, a crucial but restricted number of unit cells from the stable crystal state come together during a nucleation event. Consequently, for the nucleation of a well-formed glass, the quantity of involved molecules must significantly surpass those contributing to the structural relaxation of the glass phase, forming what is termed a CRR (Cooperative Rearranging Region) [18]. This results in a nucleation time significantly exceeding T_{eq}, the structural relaxation time. A prolonged nucleation time indicates a low likelihood of a fluctuation causing the formation of a crystal from a critical number of unit cells. This reasoning is applicable to any type of glass, spanning from traditional silicate to more contemporary metallic glasses. In recent years, there have been attempts to create metallic eyewear. These metallic solids have gained recognition due to their nanocrystalline atomic arrangement, offering numerous unique and practical features [19, 20]. Despite their admirable qualities, their limited geometry to thin ribbons and the necessity for rapid solidification at cooling rates of 10^5 K s^{-1} or higher have restricted their application for the past four decades. Recent studies, however, demonstrate that many Zr-based multi-component alloys can be cooled at significantly lower rates typically 10^2–10^3 K s^{-1} to obtain them in a glassy form. This breakthrough allows for the mass production of these alloys in their glassy state.

2.3 Glass ceramics

Crystallization of glass is typically undesirable, but when carefully controlled, it can result in the production of a superior type of polycrystalline material. Glass ceramics, defined as micro- or nanocrystalline materials derived from glasses undergoing controlled crystallization, are characterized by a composite structure comprising a crystalline phase embedded in a glassy matrix. The formation of glass ceramics is believed to involve micro- or nanoscale immiscibility in the glass, leading to amorphous phase separation [21]. The inadvertent creation of glass ceramics dates back to the 1950s when Donald Stookey made significant contributions [22]. While Stookey was not the first to explore glass ceramic materials (with French chemist Ramur [23] conducting earlier work on dense ceramic materials), his innovations spurred further studies and market introductions. Nucleating agents, crucial for finely regulating the microstructure of glass ceramics, gained recognition. Stookey's use of TiO_2 as a nucleating agent in lithium silicates, derived from his knowledge of dense thermometer opal glasses, proved highly effective, resulting in heat-resistant and mechanically robust glass ceramics. These materials found commercial success in applications such as rocket nose cones and Corning Ware® cookware [24, 25].

Apart from these uses, glass ceramics with distinctive dielectric, magnetic, and luminous characteristics have been created for diverse applications [26]. Recent research has focused on creating glass ceramics through the sintering and crystallization of glass frits, enabling the use of traditional ceramic forming techniques like slip casting and pressing. A viscous sintering process, involving the sintering of

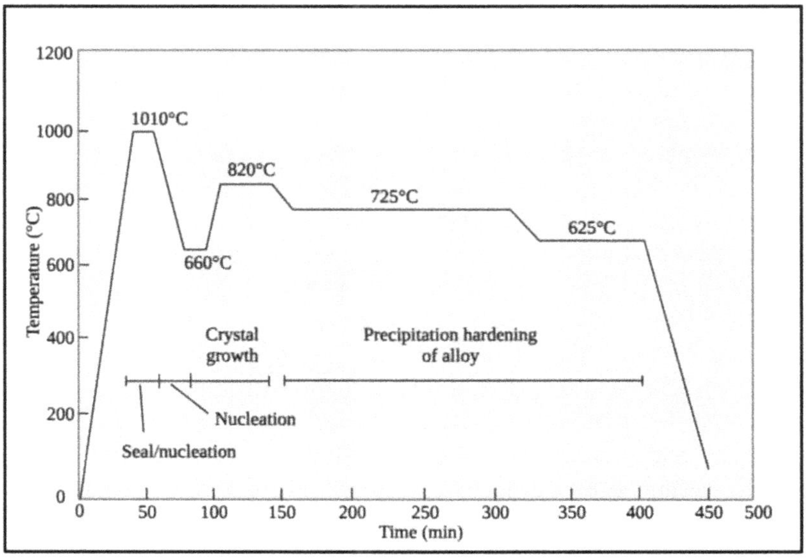

Figure 2.2. A standard thermal processing plan for inducing the crystallization of lithium aluminum silicate glasses, resulting in the formation of lithium disilicate upon sealing. Reprinted from [28], copyright (2017) with permission from Elsevier.

small-particle glass frit (3–15 μm in diameter) to full density, followed by crystallization, has been employed. This technique is utilized in the production of co-fired multi-layered substrates for electronic packaging [27].

Glass ceramic materials exhibit favorable chemical, thermal, dielectric, and biological properties, surpassing those of metals and other polymers. Their advantages over regular glasses and traditional ceramics, particularly in terms of microstructure and tunable thermo-physical qualities, make them highly desirable. The production process begins with forming the parent glass, employing methods such as melt quench, sol–gel, or CVD (Conventional Vapor Deposition). Subsequently, the glass undergoes controlled crystallization through data from techniques like DTA (Differential Thermal Analysis) and DSC (Differential Scanning Calorimetry) optimized using experimental heat treatments and carefully planned heat-treatment regimen, figure 2.2 illustrates a typical heat-treatment schedule in this glass ceramics production process.

2.3.1 Basic process of glass crystallization

Glass experiences crystallization by forming new nuclei and then allowing crystals to grow from a supercooled liquid. The application of classical theory, employing a thermodynamic approach, aids in comprehending these crystallization mechanisms. There are two clear categories of nucleation: heterogeneous nucleation, primarily happening at the interface between air and liquid or impurities, and homogeneous nucleation, occurring randomly within the entire glass matrix encompassing the crucible and setter. Once a nucleus surpasses the critical radius, it becomes observable through techniques such as x-ray diffraction or optical microscopy. The growth rate of crystals is determined by the difference in the quantity of atoms leaving the supercooled liquid and those rejoining the crystal plane. Figure 2.2 provides a fundamental depiction of how temperature impacts both the crystal growth and nucleation rate. The highest nucleation rate occurs at a temperature below the peak rate of crystal growth or near the glass transition temperature. The specific temperature dependency of the nucleation rate differs based on the type of glass. The creation of glass ceramics entails subjecting the material to heat treatment within a temperature range where nucleation and growth align, as illustrated in the shaded area of figure 2.3. This approach is taken if the objective is to actively promote crystallization.

2.3.2 Controlling the morphology of glass ceramics through diverse thermal histories

The production of glass ceramics requires precise control of heat treatment temperature and duration. The percentage of crystals in the glass, and the relationship between crystallinity, temperature for heat treatment, and duration is typically illustrated in the T–T–T diagram, as illustrated in figure 2.4. This diagram highlights crucial stages: on the left, the glass (supercooled liquid) is at 1 ppm (10^6), while on the right, crystals have the potential to develop. In the glass industry, electric furnaces are commonly used for traditional heat treatment, as discussed in section 2.3. Various heat treatment temperatures and durations are utilized to accommodate the diverse temperature preferences for nucleation and crystal development.

Advancements in laser technology, characterized by low operating costs and long life, suggest the potential for practical laser processes on glass. The rapid generation

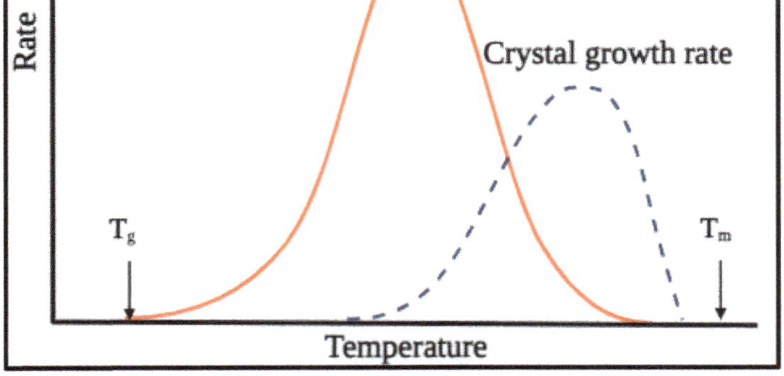

Figure 2.3. The schematic representation of effect of temperature on the rate of crystal growth and the nucleation rate.

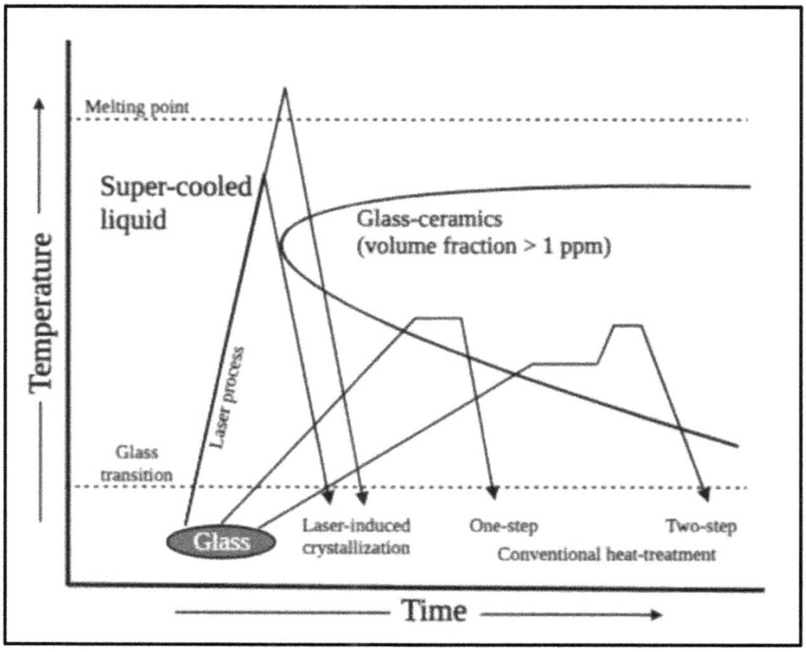

Figure 2.4. A conventional T–T–T (time–temperature–transformation) graph and the thermal treatment procedure associated with the crystallization of glass.

of a liquid phase or supercooled liquid exceeding the melting point can be achieved through laser-induced thermal gradients and localized heating. The cooling rate is regulated through the ongoing scanning of the laser beam, promoting the formation of crystals. When lithium-disilicate glass is subjected to laser irradiation, it gives rise to a one-dimensional directed crystalline pattern, exhibiting a growth rate of several tens of meters per second [29]. Research on alkali-disilicate glass crystallization has focused on the instantaneous high-temperature exposure to laser radiation, where crystal formation rates surpass those achieved through electric furnace heating [30]. Combining the heating of an electric furnace with a laser process opens up novel possibilities for the applications of glass ceramics. An intriguing portrayal centers on integrating these heat treatments into the progress of all-solid-state batteries, with the goal of addressing safety concerns associated with lithium-ion batteries. In the realm of all-solid-state batteries, establishing a robust interface between the active material and the solid electrolyte stands as a matter of paramount importance. Yamauchi and team successfully addressed this challenge through the application of traditional heat treatment, employing $Na_2FeP_2O_7$ (sodium iron phosphate) glass as the active cathode material [31, 32]. The combination of simultaneous softening flow and crystallization resulted in the formation of a resilient interface. Furthermore, Hiratsuka *et al* and the research team employed laser irradiation to create a durable interface between the solid electrolyte and the cathode made of sodium iron phosphate glass ceramic. This laser technique was applied within a temperature range conducive to prompt glass softening [33, 34]. Additionally, isothermal heat treatment in an electric furnace was employed to crystallize $Na_2FeP_2O_7$, the active material.

2.3.3 Glass ceramics for commercial use

To effectively utilize glass ceramics in practical applications, it is essential to strategically consider their advantages over other ceramic materials. Given the limitations in material design flexibility for glass ceramics, demonstrating their superiority in terms of crystal function alone can often be a challenging task. Hence, it is essential to investigate and utilize the distinct characteristics of glass ceramics, such as their uncomplicated formulation of glass procedure, distinctive microstructure, and pliable flow. This examination will explore the distinct benefits provided by conventional glass ceramic items, focusing on volumetric crystallization and surface crystallization.

2.3.4 Glass ceramics characterized by volumetric crystallization

Specially designed for applications demanding resistance to high temperatures, glass ceramics comprising $Li_2OAl_2O_3SiO_2$ have proven to be highly successful in the commercial market [35]. The utilization of a ½-quartz solid solution (s.s.) with an outstanding adverse thermal expansion property allows for the attainment of an extremely low thermal expansion coefficient (approximately $5 \times 10^{-7}\ K^{-1}$) through the process of precipitation. These items are predominantly categorized as transparent, black, and white, with the transparent variant gaining widespread popularity for its remarkable resemblance to traditional glass. The transparency is further enhanced by introducing ZrO_2 and TiO_2 as nucleation agents through ¢-quartz s.s., resulting in exceptional clarity at the submicron scale, effectively preventing light scattering [36].

Producing a microstructure in sintered ceramics requires specialized methods such as spark plasma sintering to prevent the formation of grains. Producing large-scale, highly homogeneous products at a reasonable cost poses a challenging task. Beyond its unique microstructure, the glass's rolling method provides additional advantages for various applications. The plates formed are utilized for fire-resistant windows and cooker top plates while pressing is employed to craft intricate shapes for lamp reflectors and cooking utensils [37]. Downdraw is utilized for generating tube shapes, specifically for heater covers. Overcoming the challenges associated with traditional methods is crucial, given the susceptibility of parent glass to devitrification during forming procedures. Current efforts are focused on successfully manufacturing glass ceramic sheets in substantial sizes, reflecting ongoing advancements in this field [38, 39].

2.3.5 Surface crystallization-type glass ceramics

Glass ceramics, employing surface crystallization techniques, can generate materials such as CaO_3, Al_2O_3, and SiO_2. These materials are widely employed in the construction of both interior and exterior walls [40]. The process involves the crushing of molten glass with water to form granules, leading to grain accumulation

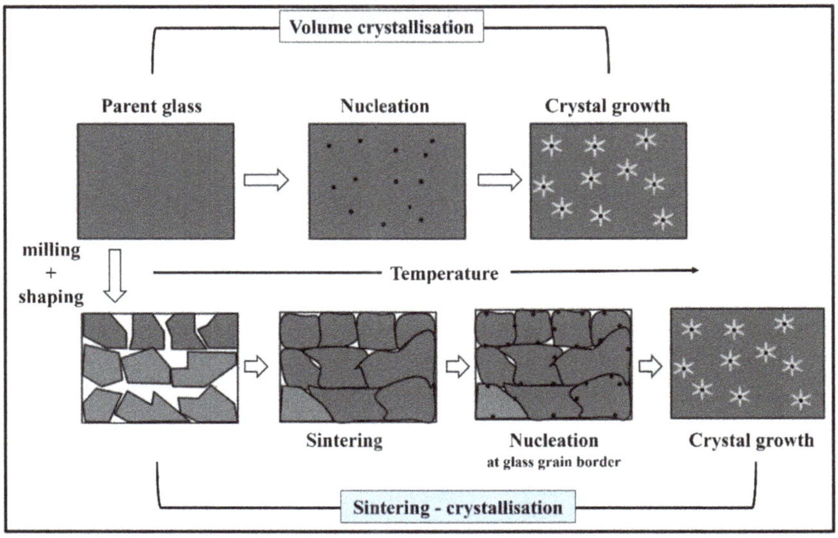

Figure 2.5. The progression of microstructure undergoes transformation from glass to glass ceramic via volume crystallization originating from either a glass or sinter-crystallization arising from a compacted glass powder.

and crystallization. Figure 2.5 illustrates the schematic diagram of this crystallization process. At approximately 850 °C, the grains fuse and smooth out, initiating wollastonite crystal precipitation at the fused particle contact around 950 °C. Crystallization is complete at 1100 °C, with the crystal growing into the interior of the grains. The matrix glass, with a crystallinity of about 30%, intricately reflects light from the outside onto the interior wollastonite crystal, exhibiting high diffuse reflectance and a distinctive aesthetic aspect. As the material comprises 70% glass phase, glass ceramics can soften and flex along the mold, taking on the required shape when placed on a metal mold and subjected to further heat treatment. In comparison to natural stone, commonly used in construction, this product showcases exceptional strength and weather resilience [41]. Glass ceramics outperform other materials due to their unique microstructure and the capacity to soften and flow, making them practical for various applications.

2.4 Characteristics of glass and glass ceramics

In this portion, we explore several attributes of glass and glass ceramics that play a crucial role in their functionality and diverse applications. Subsequently, we will delve into specific details about selected features. While glass properties are influenced by composition and manufacturing control, the functions of glass ceramics are additionally shaped by factors like microstructure and the generation of new phases. These characteristics and parameters empower them to showcase a spectrum of capabilities, spanning from simple colored glass beads to cutting-edge applications like active and biocompatible materials for implants or plasma display panels. A correlation between the properties and functions of glasses and glass ceramics is summarized in table 2.1.

Optical and optoelectronic: Pure glasses generally lack grain boundaries, preventing light scattering and ensuring isotropy, resulting in excellent transparency. Despite this, even typical windows achieve only around 80% transmission, falling short of the near-100% demand in certain technical applications [42]. Transparency is influenced by prevalent absorption processes, linked to intrinsic bonding strength or the presence of contaminants. These processes lead to transparency within specific wavelength bands, controlled by high- and low-energy photon absorption. The appealing transparency of glass becomes especially crucial when considering architectural applications such as windows for buildings, vehicles, and various kinds of optical equipment [42]. The optical flatness and smoothness of glass sheets, along with the graded refractive index profile achieved in optical fibers [43–45] during drawing, are attributed to properties like high glass melt viscosity, gravity effects, surface tension, and blowing capability. Glass ceramics, characterized by a residual glassy phase and typically non-porous, exhibit a wide range of optical and optoelectronic properties [45, 46]. Some materials, initially possessing high translucency, can be transformed into highly transparent substances by reducing crystallite size to the nanoscale. Moreover, adjusting the kind and morphology of the crystalline phase allows the creation of highly opaque materials in various colors. For specialized applications like laser action, rare-earth elements such as Nd can be doped into glasses and low-temperature glass ceramics [47–52]. Another noteworthy characteristic of many of these materials is photosensitivity, contributing to various optoelectronic phenomena.

Thermal: The thermal expansion coefficient (TEC) stands out as a crucial thermal parameter in numerous applications, with a special focus on the encapsulation and sealing of telescopic mirrors. By carefully manipulating the composition and morphology of crystalline phases, significant control can be achieved in various areas such as glazing and cookware [53, 54]. Glass ceramics exhibit the unique capability of achieving a spectrum of TECs ranging from negative to positive, even through zero. This is in stark contrast to conventional ceramics, where concerns about shrinking prevail; glass ceramics, on the other hand, experience minimal shrinking.

Mechanical: Glass is a brittle material. However, as mentioned earlier, transforming it into glass ceramics considerably improves the mechanical strength [55]. If one were to consider simply the tensile stress for failure (σ_{th}), it can be expressed as

$$\sigma_{\text{th}} = \sqrt{\frac{E\zeta}{r_o}}$$

In the given context, E represents Young's modulus, ζ denotes surface energy, and the Si–O bond's length is denoted by r_o= 1.5 Å. With $E = 70$ GPa and $\zeta = 0.6\ \text{J m}^{-2}$, the calculated σ_{th} is approximately 16 GPa. However, experimental findings impose a 50 MPa upper limit on the strength of flat glass sheets [55]. This implies that the strength of glasses can be reduced below predicted levels due to surface defects. As a

Table 2.1. Correlation between property and function of glasses and glass ceramics.

Property category	Property	Function
Physical	Density	A fundamental property influences directly the refractive index and other properties.
Optical	Refractive index Dispersion Transmission and absorption	Control light path. Control chromatic aberration, improve image quality. Control transparency window of undoped glasses and selective absorption of doped (3d and 4f metal ions, nanometal, semiconductor, etc) glasses from ultraviolet (UV) to far infrared (FIR) regions of the electromagnetic spectrum.
Photonic	Nonlinear optical (NLO) Property Photoluminescence and emission Magneto-optical property	Optical switching and modulation. Absorption, luminescence, and lasing action. Control of optical effect under magnetic field.
Thermal	Coefficient of thermal expansion Thermal conductivity	Control of dimensional changes with temperature. Transport of heat energy through the material.
Electrical Electronic	Electrical conductivity Dielectric property: permittivity (dielectric constant) and dielectric loss Electronic conduction	Resistance to mobility of ions or free electrons. Extent of polarizability of ions and loss of electrical energy under electric field. Conduction of electrical charge by electrons (e7) or holes (h+)
Microstructure	Morphology (size and shape) and orientation of grain	Control properties of the glasses and glass ceramics.
Mechanical	Elastic property: Young's and bulk moduli Hardness: Knoop and Vickers	Control breakage of the glass. Control strength of the glass.
Chemical	Fracture toughness Durability	Resistance under load. Inertness to environment or chemicals.

result, significant efforts are being directed towards minimizing surface damage in glass products. The exceptional strength of glass fibers is harnessed in glass-reinforced plastic (GFRPS), a material with high strength-to-weight and stiffness-to-weight ratios [56]. Various techniques, such as tempering, have been developed to enhance the durability of glasses. Glass ceramics demonstrate enhanced mechanical characteristics in comparison to conventional ceramics, primarily attributed to their near-total absence of porosity. Despite not matching the flexural strength of metal alloys, glass ceramics can achieve flexural strengths of 500 MPa. Careful micro-structure tuning allows for accepting critical fracture propagation load (K1c) values greater than 3 MPa · m 0.5. Recent advancements in processing have increased the durability of metallic glasses based on Ti and Zr, surpassing that of steel [57]. Additionally, certain glass ceramic materials [54, 58–60] can be machined, enabling their formation using conventional metalworking equipment [60].

Electrical and magnetic: Glasses without metallic elements exhibit semiconductor or insulator properties at room temperature, determined by their structural arrangement, bonding, and composition. Some of these glasses demonstrate ionic conductivity at elevated temperatures, with the potential for achieving even faster ionic conduction comparable to crystalline ceramics. This ionic conduction also gives rise to mixed alkali effects in glasses. Glass ceramics with high resistivity find utility as isolators, spacers, or insulators, while in microwave applications, substrates with low dielectric loss are preferred over pure alumina [61]. The notable ionic conductivity of specific glass ceramics has led to their development as electrolytes for battery applications [62, 63].

Biological: Extensive research attention and endeavors are dedicated to exploring the application of glass and glass ceramics in the biomedical field [64, 65]. A bioactive glass ceramic, fostering the development of a physiologically active hydroxyapatite layer on its surface, facilitates effective bonding with both soft tissues and bone [66, 67]. Specific requirements include considerations such as Young's modulus, the incorporation of suitable doping for flexibility, and processing techniques to showcase magnetic characteristics and biocompatibility in certain instances [68, 69, 71, 72]. Recent studies have indicated that nanostructured materials could significantly influence biomedical applications [73, 74].

Chemical: The inherent structure of glasses imparts robustness and chemical inertness to them. In aqueous or acidic environments, the majority of silicate glasses display minimal deterioration and exhibit high durability [75, 76]. The absence of grain boundaries is advantageous since many materials corrode primarily at these boundaries. Consequently, glasses are commonly employed to contain various liquids. While degradation is seldom an issue in diverse applications, it is crucial to acknowledge that glasses containing significant proportions of alkali or alkaline ions may not be entirely 'inert,' as these cations can potentially leach out [77]. Nonetheless, depending on the intended purpose, glass ceramics can be tailored to possess characteristics ranging from high stability to resorbability [70]. The optimization of the crystalline phases' form and nature, along with the interface between the crystalline and glassy phases, plays a pivotal role in regulating the chemical durability qualities of glass ceramics.

Environmental compatibility and ease of recycling: While polymers have supplanted glasses in various traditional applications [78], glasses still offer substantial recycling advantages. Glass can be endlessly reused and is nearly fully recyclable. Additionally, in certain instances, glass containers are preferred for their aesthetic appeal, especially in storing perfumes and alcohol.

2.5 Challenges

Glass ceramics exhibit enhanced toughness due to their combination of crystalline and glassy phases. Beall [79] highlighted the discovery of an exceptionally durable glass ceramic in 1991, featuring precipitated chain silicate crystals, and drew comparisons to the microstructure of jade, a well-known naturally resilient stone. Fu [80] classified common silicate glass ceramics into three categories: aluminosilicates, fluosilicates, and silicates, all exhibiting remarkable toughness and strength. Table 2.2 illustrates instances of glass ceramic compositions, along with their corresponding flexural strength values and fracture toughness (KIC) [81, 84]. Recent research has focused on high-toughness glass ceramics, such as the Li_2O–SiO_2 system and Leucite ($KAlSi_2O_6$) glass ceramics, for applications like artificial dental crowns [82, 85]. Glass ceramics with precipitated canasite or enstatite as their primary phases characterized by a crystalline structure display superior fracture toughness. Glass ceramics composed of enstatite showcase impressive toughness, credited to crater slippage and deflection along interlocking twinned crystals [79]. Meanwhile, canasite glass ceramics achieve heightened fracture toughness through mechanisms such as high energy absorption via crack branching, residual strain resulting from thermal expansion anisotropy, and deflection along cleavage planes [79]. Toughening mechanisms, microcrack, crack bridging, and including phase transformation, have been proposed since the study of toughened ZrO_2 ceramics in 1975 [86]. Remaining stress from differing rates of thermal expansion between crystals and residual glass is significant in glass ceramics [87, 88]. Seavena and colleagues extensively investigated the high toughness of Li_2O–SiO_2 model glass ceramics, considering fracture deflection, break bowing, crack trapping, and crack bridging [89]. While toughness generally increases with crystallinity, Inage *et al* [90] observed the highest fracture toughness in CaO–Al_2O_3–SiO_2 (CAS) glass ceramics at 20%–40% crystallinity, declining with further crystallization. The hexagonal metastable $CaAl_2Si_2O_8$ crystals in CAS glass ceramics exhibit card house-like structures, contributing to microcrack formation at approximately 20 weight percent crystallinity (figure 2.5) [32]. Maeda and colleagues investigated the *R*-curve behavior during fracture in CAS glass ceramics, noting similarities to mica-based glass ceramics [91, 92]. The exceptional durability of these glass ceramics is ascribed to the facile creation of microcracks, which results from the robust cleavage of the hexagonal $CaAl_2Si_2O_8$ crystal and mica. X-ray micro-CT is employed to understand how microcrack formation influences toughening [93]. Although light scattering from precipitated crystals often leads to a decrease in transparency. The quest for a novel material combining high toughness and transparency remains a challenging endeavor (figure 2.6).

2.6 Conclusion

A substance lacking a defined structure over a limited atomic range is termed a glass, characterized by short-range order. The attributes of glass are influenced by the melt type, processing methods, and the inclusion of additives for diverse applications. Consequently, a broad spectrum of glasses is now available for various innovative

Table 2.2. Bending strength and fracture toughness of typical silicate crystallized glasses.

Glass ceramic family	Primary crystal structure	Flexural strength, MPa	K_{IC}, MPa · $m^{1/2}$ (method)	Precursor glass compositions (wt %)	References
Aluminosilicate	β-Spodumene	100–150	1.0 (CNSB)	65.1 SiO_2—20.1 Al_2O_3—2.0 B_2O_3—3.6 Li_2O—0.4 Na_2O—1.8 MgO—2.2 ZnO—4.4 TiO2—0.4 SnO_2	[80]
	Leucite (Ivoclar IPS Empress®)	110–185	1.3 (SEVNB)	59–63 SiO_2—19–23.5 Al_2O_3—10–14 K_2O—3.5–6.5 Na_2O—0–1 B_2O_3—0–1 CeO_2—0.5–3 CaO—0–1.5 BaO—0–0.5 TiO_2	[82]
	Cordierite (Corning 9606)	200	2.2 (CNSB)	56.1 SiO_2—19.8 Al_2O_3—14.7 MgO—8.9 TiO_2—0.3 As_2O_3—0.1 CaO—0.1 Fe_2O_3	[80]
	Hexagonal $CaAl_2Si_2O_8$	103	2.2 (SEVNB)	55 SiO_2—20 Al_2O_3—25 CaO	[83, 84]
Fluorosilicate	Mica (Macor)	94	1.53 (SEVNB)	47.2 SiO_2—8.5 B_2O_3—16.7 Al_2O_3—14.5 MgO—9.5 K_2O—6.3 F	[80]
	Canasite	300	5.0 (CNSB)	57 SiO_2—2 Al_2O_3—11 CaO—13 CaF—8 Na_2O—9 K_2O	[80]
	F-K-Richiterite	220	3.2 (CNSB)	67.1 SiO_2—1.8 Al_2O_3—14.2 MgO—4.7 CaO—3.0 Na_2O—4.8 K_2O—0.75 Li_2O—0.3 BaO—1.0 P_2O_5—0.2 Sb_2O_3—3.5 F	[80]
Silicate	Lithium Disilicate	217	3.3 (SENB)	74.2 SiO2—3.54 Al2O3—15.4 Li_2O—3.25 K_2O—3.37 P_2O_5	[80]
	Enstatite	200	4.6 (CNSB)	54 SiO_2—33 MgO—13 ZrO_2	[80]
	Enstatite + Spinel	107	1.3 (CNSB)	47.1 SiO_2—22.1 Al_2O_3—16.9 MgO—1.7 ZnO—12.3 TiO_2	[81]

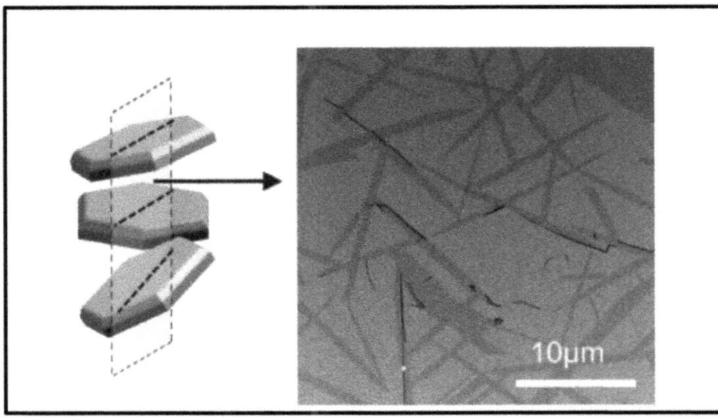

Figure 2.6. Scanning electron microscopy (SEM) visualization depicting the propagation of microcracks along the hexagonal $CaAl_2Si_2O_8$ crystals within CAS glass ceramics. Reprinted from [32] John Wiley & Sons, copyright (2019).

purposes. The deliberate crystallization of glasses, shaping a tailored microstructure, and adjusting the crystalline to glassy phase ratio, have revolutionized glass-to-glass ceramics, unlocking a multitude of enhanced capabilities. Glasses and glass ceramics, derived from oxides, chalcogenides, fluorides, polymers, and other materials, exhibit properties akin to metals, semiconductors, and insulators. Depending on their kind, structure, and the quantity of crystalline phases, composite glass ceramic materials can manifest a diverse range of advantageous qualities. The capacity for nanocrystallization has paved the way for inventive applications in memory storage and optics. A deeper understanding of the relationships between the structure and properties of glasses holds the potential for novel applications in energy storage, construction, structural engineering, biology, and various other domains. Despite facing competition from numerous alternatives, glasses remain indispensable functional materials. They occupy specific niches in the market where their suitability is exceptional, rendering them challenging to substitute.

References

[1] El-Meliegy E and Van Noort R 2011 *Glasses and Glass Ceramics for Medical Applications* (Berlin: Springer Science & Business Media)

[2] Sakamoto A and Yamamoto S 2010 Glass–ceramics: engineering principles and applications *Int. J. Appl. Glass Sci.* **1** 237–47

[3] Karasu B, Bereket O, Biryan E and Sanoğlu D 2017 The latest developments in glass science and technology *El-Cezeri* **4** 209–33

[4] Turner W E 1961 *Ancient Glass and Glass-Making* (London: Proceedings of the Chemical Society)

[5] Doremus R H 1991 Glass: an overview *Concise Encyclopedia of Advanced Ceramic Materials* (Elsevier) pp 166–70

[6] Chopinet M H 2019 The history of glass *Springer Handbook of Glass* (Springer) pp 1–47

[7] Stookey S D 1953 Chemical machining of photosensitive glass *Ind. Eng. Chem.* **45** 115–8

[8] Richet P 2021 A history of glass science *Encyclopedia of Glass Science, Technology, History, and Culture* **2** 1413–40

[9] Sugarman B 1967 Strength of glass (a review) *J. Mater. Sci.* **2** 275–83

[10] Pollington S and van Noort R 2009 An update of ceramics in dentistry *Int. J. Clin. Dent.* **2** 283–307

[11] Cuce E and Riffat S B 2015 A state-of-the-art review on innovative glazing technologies *Renew. Sustain. Energy Rev.* **41** 695–714

[12] Siebers F, Weiss E and Gabel F 2015 Inventors; Schott AG, assignee. Glass ceramic as a cooktop for induction heating having improved colored display capability and heat shielding, method for producing such a cooktop, and use of such a cooktop *United States Patent US* 9,018,113

[13] Fernandes H R, Gaddam A, Rebelo A, Brazete D, Stan G E and Ferreira J M 2018 Bioactive glasses and glass-ceramics for healthcare applications in bone regeneration and tissue engineering *Materials* **11** 2530

[14] Wang W C, Zhou B, Xu S H, Yang Z M and Zhang Q Y 2019 Recent advances in soft optical glass fiber and fiber lasers *Prog. Mater Sci.* **101** 90–171

[15] Mouritz A P 2012 *Introduction to Aerospace Materials* (Amsterdam: Elsevier)

[16] Leuzzi L and Nieuwenhuizen T M 2007 *Thermodynamics of the Glassy State* (Boca Raton, FL: CRC Press)

[17] Kingery W D, Bowen H K and Uhlmann D 1976 *Introduction to Ceramics* (New York: Wiley)

[18] Adam G and Gibbs J H 1965 On the temperature dependence of cooperative relaxation properties in glass-forming liquids *J. Chem. Phys.* **43** 139–46

[19] Inoue A 1998 *Bulk Amorphous Alloys: Preparation and Fundamental Characteristics* (Trans Tech) pp 1–24

[20] Inoue A 2000 Stabilization of metallic supercooled liquid and bulk amorphous alloys *Acta Mater.* **48** 279–306

[21] Pinckney L R and Beall G H 2008 Microstructural evolution in some silicate glass–ceramics: a review *J. Am. Ceram. Soc.* **91** 773–9

[22] Doherty P E, Lee D W and Davis R S 1967 Direct observation of the crystallization of Li_2O-Al_2O_3-SiO_2 glasses containing TiO_2 *J. Am. Ceram. Soc.* **50** 77–81

[23] Dawson V P 1987 *Nature's Enigma: The Problem of the Polyp in the Letters of Bonnet, Trembley and Réaumur* (Philadelphia, PA: American Philosophical Society)

[24] Kothiyal G P, Ananthanarayanan A and Dey G K 2012 *Glass and Glass-Ceramics* (London: Elsevier)

[25] Stookey S D 2000 *Explorations in Glass: An Autobiography* (New York: Wiley-Blackwell)

[26] Beall G H 1992 Design and properties of glass-ceramics *Annu. Rev. Mater. Sci.* **22** 91–119

[27] Tummala R R 1991 Ceramic and glass-ceramic packaging in the 1990s *J. Am. Ceram. Soc.* **74** 895–908

[28] Sharma N and Dalvi A 2019 Insertion of binary LiCl-P_2O_5 glass between Li+ NASICON crystallites and its effect on controlling inter-grain transport *Solid State Ionics* **342** 115082

[29] Shinozaki K, Hashimoto K, Honma T and Komatsu T 2015 TEM analysis for crystal structure of metastable BiBO3 (II) phase formed in glass by laser-induced crystallization *J. Eur. Ceram. Soc.* **35** 2541–6

[30] Lu Z P and Liu C T 2002 A new glass-forming ability criterion for bulk metallic glasses *Acta Mater.* **50** 3501–12

[31] Sato F, Honma T, Komatsu T, Shinozaki K, Ina T and Yamauchi H 2022 Formation of highly dispersed tin nanoparticles in amorphous silicates for sodium ion battery anode *J. Phys. Chem. Solids* **161** 110377
[32] Li C, Zhang G, Zheng H, Zhang F and Liu K 2023 Study on glass-ceramic prepared by zinc leaching residue and solidification mechanism of heavy metals *J. Clean. Prod.* **426** 139021
[33] Fraser R and Girtan M 2023 A selective review of ceramic, glass and glass–ceramic protective coatings: general properties and specific characteristics for solar cell applications *Materials* **16** 3906
[34] Hiratsuka M, Honma T and Komatsu T 2021 Vitrification of maricite $NaFePO_4$ crystal by laser irradiation and enhanced sodium ion battery performance *J. Alloys Compd.* **885** 160928
[35] Jiang Q K *et al* 2008 Zr–(Cu, Ag)–Al bulk metallic glasses *Acta Mater.* **56** 1785–96
[36] Khmyrov R S, Podrabinnik P A, Tarasova T V, Gridnev M A, Korotkov A D, Grigoriev S N, Kurmysheva A Y, Kovalev O B and Gusarov A V 2023 Partial crystallization in a Zr-based bulk metallic glass in selective laser melting *Int. J. Adv. Manuf. Technol.* **126** 5613–31
[37] Rebouças L B, Souza M T, Raupp-Pereira F and Novaes de Oliveira A P 2019 Characterization of Li_2O-Al_2O_3-SiO_2 glass-ceramics produced from a Brazilian spodumene concentrate *Cerâmica* **65** 366–77
[38] Kim S W, Lee H S, Jun D S, Lee S E, Lee J H and Lee H C 2023 Enhancing the plasma-resistance properties of Li_2O–Al_2O_3–SiO_2 glasses for the semiconductor etch process via alkaline earth oxide incorporation *Materials* **16** 5112
[39] Fujita S and Tanabe S 2015 Glass-seramics and solid-state lighting *Appl. Glass Sci.* **6** 356–63
[40] Mandal S, Chatterjee R, Nag S, Manna S, Jana S, Biswas K and Ambade B 2023 Low expansion glass-ceramics using industrial waste and low-cost aluminosilicate minerals: fabrication and characterizations *T. Indian Ceram. Soc.* **82** 46–55
[41] Salamon M B and Jaime M 2001 The physics of manganites: structure and transport *Rev. Mod. Phys.* **73** 583
[42] Bembnowicz P and Golonka L J 2010 Integration of transparent glass window with LTCC technology for μTAS application *J. Eur. Ceram. Soc.* **30** 743–9
[43] Tanabe S 2002 Rare-earth-doped glasses for fiber amplifiers in broadband telecommunication *C.R. Chim.* **5** 815–24
[44] Paul M C, Sen R, Youngman R E and Dhar A 2008 Fluorine incorporation in silica glass by the MCVD process: study of fluorine incorporation zone, evaluation of optical properties and structure of the glass *J. Non-Cryst. Solids* **354** 5408–20
[45] Gonçalves M C, Santos L F and Almeida R M 2002 Rare-earth-doped transparent glass ceramics *C.R. Chim.* **5** 845–54
[46] Hou Z, Su C, Zhang Y, Zhang H, Zhang H, Shao J and Meng Q 2006 Effect of crystallization of Li_2O-Al_2O_3-SiO_2 glasses on luminescence properties of Nd^{3+} ions *J. Rare Earths* **24** 418–22
[47] Kravchenko V B and Rudnitskiĭ Y P 1979 Phosphate laser glasses *Sov. J. Quantum Electron.* **9** 399
[48] Campbell J H, Suratwala T I, Thorsness C B, Hayden J S, Thorne A J, Cimino J M, Marker Iii A J, Takeuchi K, Smolley M and Ficini-Dorn G F 2000 Continuous melting of phosphate laser glasses *J. Non-Cryst. Solids* **263** 342–57
[49] Campbell J H and Suratwala T I 2000 Nd-doped phosphate glasses for high-energy/high-peak-power lasers *J. Non-Cryst. Solids* **263** 318–41

[50] Campbell J H, Wallerstein E P, Toratani H, Meissner H E, Nakajima S and Izumitani S 1995 Effects of process gas environment on platinum-inclusion density and dissolution rate in phosphate laser glasses *Glass Sci. Technol. (Frankfurt)* **68** 59–69

[51] Koechner W 2013 *Solid-State Laser Engineering* (Berlin: Springer)

[52] Ehrmann P R, Carlson K, Campbell J H, Click C A and Brow R K 2004 Neodymium fluorescence quenching by hydroxyl groups in phosphate laser glasses *J. Non-Cryst. Solids* **349** 105–14

[53] Koepke B G and Stokes R J 1979 Grinding forces in polycrystalline ceramics. NBS Special Publication **562** 75

[54] Taruta S, Sakata M, Yamaguchi T and Kitajima K 2008 Crystallization process and some properties of novel transparent machinable calcium-mica glass-ceramics *Ceram. Int.* **34** 75–9

[55] Aben H, Guillemet C, Aben H and Guillemet C 1993 Basic photoelasticity *Photoelasticity of Glass* (Springer) pp 51–68

[56] Barre S, Chotard T and Benzeggagh M L 1996 Comparative study of strain rate effects on mechanical properties of glass fibre-reinforced thermoset matrix composite *Compos. Part A: Appl. Sci. Manuf.* **27** 1169–81

[57] Launey M E, Hofmann D C, Johnson W L and Ritchie R O 2009 Solution to the problem of the poor cyclic fatigue resistance of bulk metallic glasses *Proc. Natl Acad. Sci.* **106** 4986–91

[58] KhatibZadeh S, Samedani M, Yekta B E and Hasheminia S 2008 Effect of sintering and melt casting methods on properties of a machinable fluor-phlogopite glass–ceramic *J. Mater. Process. Technol.* **203** 113–6

[59] Grossman D G 1972 Machinable glass-ceramics based on tetrasilicic mica *J. Am. Ceram. Soc.* **55** 446–9

[60] Lianjie M and Aibing Y 2007 Influencing of technological parameter on tools wear during turning fluorophlogopite glass-ceramics *J. Rare Earths* **25** 330–3

[61] Hirano S I, Hayashi T and Hattori A 1991 Chemical processing and microwave characteristics of (Zr, Sn) TiO_4 microwave dielectrics *J. Am. Ceram. Soc.* **74** 1320–4

[62] Xu X, Wen Z, Gu Z, Xu X and Lin Z 2004 Lithium ion conductive glass ceramics in the system $Li_{1.4}Al_{0.4}(Ge_{1-x}Ti_x)_{1.6}(PO_4)_3$ (x= 0–1.0) *Solid State Ionics* **171** 207–13

[63] Cruz A M, Ferreira E B and Rodrigues A C 2009 Controlled crystallization and ionic conductivity of a nanostructured $LiAlGePO_4$ glass–ceramic *J. Non-Cryst. Solids* **355** 2295–301

[64] Vallet-Regi M and González-Calbet J M 2004 Calcium phosphates as substitution of bone tissues *Prog. Solid State Chem.* **32** 1–31

[65] Hench L L 2001 Bioglass and similar materials *Encyclopedia of Materials: Science and Technology* (Elsevier) pp 563–8

[66] Hu J, Liu X and Ma P X 2008 Biomineralization and bone regeneration *Principles of Regenerative Medicine* (Amsterdam: Elsevier) pp 744–55

[67] Goller G, Demirkıran H, Oktar F N and Demirkesen E 2003 Processing and characterization of bioglass reinforced hydroxyapatite composites *Ceram. Int.* **29** 721–4

[68] Wust P, Hildebrandt B, Sreenivasa G, Rau B, Gellermann J, Riess H, Felix R and Schlag P M 2002 Hyperthermia in combined treatment of cancer *Lancet Oncol.* **3** 487–97

[69] Ebisawa Y, Miyaji F, Kokubo T, Ohura K and Nakamura T 1997 Bioactivity of ferrimagnetic glass-ceramics in the system FeO_2 $Fe2O3_2$ CaO_2 SiO_2 *Biomaterials* **18** 1277–84

[70] Hench L L 2006 The story of Bioglass® *J. Mater. Sci., Mater. Med.* **17** 967–78

[71] Peitl O, Zanotto E D and Hench L L 2001 Highly bioactive P_2O_5–Na_2O–CaO–SiO_2 glass-ceramics *J. Non-Cryst. Solids* **292** 115–26

[72] Singh R K, Srinivasan A and Kothiyal G P 2009 Evaluation of $CaO–SiO_2–P_2O_5–Na_2O–Fe_2O_3$ bioglass-ceramics for hyperthermia application *J. Mater. Sci., Mater. Med.* **20** 147–51
[73] Heness G and Ben-Nissan B 2004 Innovative bioceramics *Mater. Forum* **27** 104–14
[74] Ying J Y 2009 Nanobiomaterials *Nano Today* **4** 1–2
[75] Brown J T and Kobayashi H 1998 Is your glass full of water?—Part II *A Collection of Papers Presented at the 58th Conference on Glass Problems: Ceramic Engineering and Science Proceedings* **vol 19** (Hoboken, NJ: Wiley) pp 1–13
[76] Tadjiev D R and Hand R J 2010 Surface hydration and nanoindentation of silicate glasses *J. Non-Cryst. Solids* **356** 102–8
[77] Cannillo V, Pierli F, Ronchetti I, Siligardi C and Zaffe D 2009 Chemical durability and microstructural analysis of glasses soaked in water and in biological fluids *Ceram. Int.* **35** 2853–69
[78] Alwaeli M 2010 The impact of product charges and EU directives on the level of packaging waste recycling in Poland *Resour. Conserv. Recycl.* **54** 609–14
[79] Beall G H 1991 Chain silicate glass-ceramics *J. Non-Cryst. Solid.* **129** 163–73
[80] Gupta P K 1996 Non-crystalline solids: glasses and amorphous solids *J. Non-Cryst. Solids* **195** 158–64
[81] Holand W and Beall G H 2019 *Glass-Ceramic Technology* (New York: Wiley)
[82] Honma T, Maeda K, Nakane S and Shinozaki K 2022 Unique properties and potential of glass-ceramics *J. Ceram. Soc. Jpn.* **130** 545–51
[83] Maeda K, Akatsuka K and Yasumori A 2021 Practical strength of damage-resistant $CaO–Al_2O_3–SiO_2$ glass-ceramic *Ceram. Int.* **47** 8728–31
[84] Soares V O, Serbena F C, dos Santos Oliveira G, da Cruz C, Muniz R F and Zanotto E D 2021 Highly translucent nanostructured glass-ceramic *Ceram. Int.* **47** 4707–14
[85] Bamba N 1986 Effects of nano-sized silicon carbide *J. Mater. Sci. Lett.* **5** 159–62
[86] Evans A G 1990 Perspective on the development of high-toughness ceramics *J. Am. Ceram. Soc.* **73** 187–206
[87] Fu L S, Sheu Y C, Co C M, Zhong W F and Shen H D 1985 Fundamentals of Microcrack Nucleation Mechanics *NASA Contractor Report 3851* (Washington, DC: NASA)
[88] Theunissen G S, Winnubst A J and Burggraaf A J 1993 Sintering kinetics and microstructure development of nanoscale Y-TZP ceramics *J. Eur. Ceram. Soc.* **11** 315–24
[89] Bohmer M and Almond E A 1988 Mechanical properties and wear resistance of a whisker-reinforced zirconia-toughened alumina *Mater. Sci. Eng.* A **105–106** 105–16
[90] Inage K, Akatsuka K, Iwasaki K, Nakanishi T, Maeda K and Yasumori A 2020 Effect of crystallinity and microstructure on mechanical properties of $CaO\text{-}Al_2O_3\text{-}SiO_2$ glass toughened by precipitation of hexagonal $CaAl_2Si_2O_8$ crystals *J. Non-Cryst. Solids.* **534** 119948
[91] Becher P F 1986 Toughening behavior in ceramics associated with the transformation of tetragonal ZrO_2 *Acta Metall.* **34** 1885–91
[92] Serbena F C, Mathias I, Foerster C E and Zanotto E D 2015 Crystallization toughening of a model glass-ceramic *Acta Mater.* **86** 216–28
[93] So-Won K, Hwan-Seok L, Jun D S, Lee S E and Joung-Ho L 2023 Enhancing the plasma-resistance properties of $Li_2O–Al_2O_3–SiO_2$ glasses for the semiconductor etch process via alkaline earth oxide incorporation *Materials* **16** 5112
[94] Mazzei A C and Rodrigues J A 2000 Alumina-mullite-zirconia composites obtained by reaction sintering: part I. Microstructure and mechanical behaviour *J. Mater. Sci.* **35** 2807–14

IOP Publishing

Glass-based Materials
Advances in energy, environment and health
Sathish-Kumar Kamaraj and Arun Thirumurugan

Chapter 3

Advancement of optical fibre in the medical field

Chandraraj Shanmuga Sundari and Naushad Edayadulla

3.1 Introduction

The evolution of systems and our knowledge of the interactions between light and matter have been made possible by optics, from x-rays to optical endomicroscopy devices. The study of how light interacts with biological materials, such as tissue, and its applications in biomedical research, diagnosis, treatment, monitoring, imaging, and surgery is known as biomedical optics. It is sometimes referred to as biophotonics and medical optics. The area studies biological systems at the molecular, cellular, and tissue levels by fusing the concepts of biology, photonics, and optics. The biomedical fields include: (a) biological element imaging techniques, (ranging from cells to organs); (b) non-invasive biometric parameters, such as blood O_2 and glucose monitoring; (c) light-built therapy for injured or diseased tissue; (d) diagnosis of diseased or injured tissue; (e) tracking the development of wound healing; and (f) surgical techniques, such as laser cutting, tissue removal and tissue ablation.

In order to examine the mechanical, functional, structural, chemical, and biological aspects of biological material and systems, the employment of various types of optical fibres is appealing for biophotonic applications. This is because fibres enable the precise lighting of tissue sections. Furthermore, biophotonic techniques which are widely employed to study and track human health and wellbeing also heavily rely on optical fibres. Biophotonic uses of optical fibres include laser surgery, dentistry, dermatology, spectroscopy, blood parameters monitoring, biostimulation, imaging, and health status monitoring. The fibres transfer optical power levels of less than 1 W in biological photonics applications including imaging and fluorescence spectroscopy. The fibres must transfer optical power levels of 10 W or more in other applications. One main use is in laser surgery, which also covers cardiovascular operations, dentistry, ophthalmology, dermatology, and oncological therapies. Other applications include bone ablation.

doi:10.1088/978-0-7503-5904-7ch3

Fibre optics are safe from electromagnetic interference, and they can also endure extremely elevated temperatures, powerful electromagnetic fields like MRIs, and ionizing radiation, making them the ideal medical tool. The use of optical fibres in and around the human body is excellent since they are nontoxic, chemically inert, and fundamentally safe. The imaging and illumination parts of endoscopes are among the most common places where fibre optics is used in medicine. Optical coherence tomography (OCT), a form of biomedical sensor, is a medical imaging technique that makes use of imaging sensors to collect micron-scale, three-dimensional images from inside an optically scattering medium, such as biological tissue. Due to the fibre optic's remarkable mechanical flexibility and small size, it is possible to access bodily parts that would otherwise be inaccessible. The next generation of smartphone-compatible optical devices is being driven by new developments in optics. Given that cellphones can transfer data through mobile communication channels, this breakthrough is helpful for scientists and medical professionals conducting study in remote areas. When compared to reusable flexible catheters, many single-use fibre-optic surgical catheters have shown to operate satisfactorily. From the surgeon's headlamp to sophisticated robotic-assisted surgery systems and endoscopes, fibre-optic technology has significantly improved medical practice and revolutionized invasive surgery. Optical fibres have found niche uses in sensing because of their special qualities, which include their tiny size, ability to detect electromagnetic radiation without interference, and potential for distant sensing. Figure 3.1 describes the level of maturity of different fibre optic sensor systems.

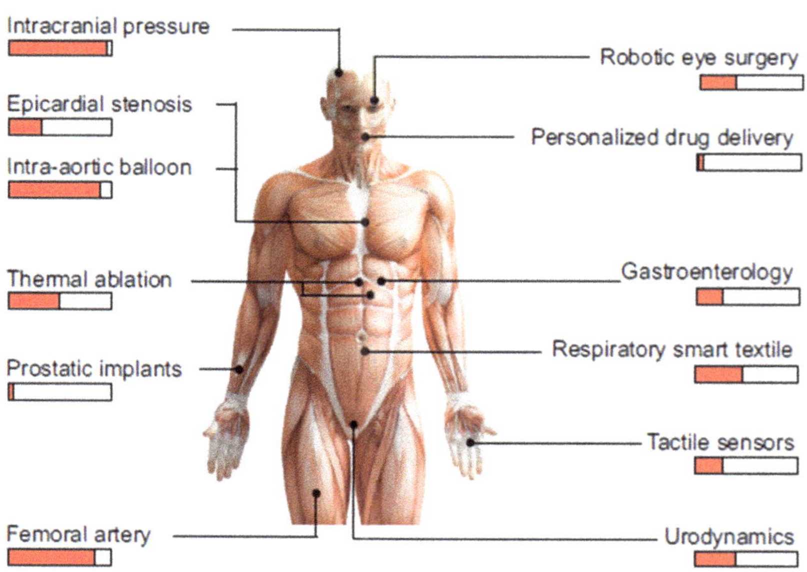

Figure 3.1. Fibre optic sensor systems for different biomedical applications. Reprinted from [1], Copyright (2024), with permission from Elsevier.

3.2 Optical coherence tomogram

OCT technology has been utilized to quantify corneal thickness and the axial length of both normal and cataract-ridden eyes *in vivo*. By using optical tomography on ocular tissues, Fercher *et al* [2] were able to generate an *in vivo* OCT of a portion of a human ocular fundus at the optical disk. As a light source, a superluminescent diode with wavelength A of 830 nm was employed. By producing distinct interferometric signals at angle increments of 0.5°, or length increments of around 160 μm, they were able to capture the image. OCT was first shown as a novel diagnostic method for non-invasive, cross-sectional imaging of retinal structure with micrometre resolution by Swanson *et al* in 1993 [3]. The advancements allow for real-time picture updating on a computer monitor and image acquisition in live patients in less than 3 s. In 1994, Izatt *et al* [4] employed OCT as a diagnostic instrument for noncontact biometry, anterior chamber angle evaluation, intraocular mass-tumour diagnosis and monitoring, and clarification of anomalies related to the cornea, iris, and crystalline lens.

In 1996, Carl Zeiss brought a cutting-edge OCT to market after that it became a widely used imaging technique in clinical detection. Intravascular imaging and highly dispersed tissues like skin and teeth may now be imaged with OCT by utilizing a broad-spectrum light source. Colston and co-workers first employed this system for intraoral imaging of human dental tissues by combining the transverse scanning optics and the interferometer sample arm. For both hard and soft tissues, the system's typical imaging depth ranged from 3 to 1.5 mm. They went over the use of this imaging technology in dentistry and showed how it may be used to diagnose periodontal disease, find cavities, and assess dental restorations [5].

It has been used to diagnose many skin and eye conditions. For the purpose of imaging dental structures cross-sectionally, this approach has been adjusted. In order to assess the technique's potential for diagnostic use in dentistry, Baumgartner *et al* [6] used it to acquire tomographic pictures of decaying human teeth. Phase retardation pictures obtained with polarization-sensitive OCT and classical OCT images derived from reflectivity measurements were captured. It was shown that polarization-sensitive OCT can offer further data, most likely pertaining to the dental material's scattering characteristics and/or mineralization state. OCT's ability to employ near-infrared light rather than ionizing radiation is one of its appealing qualities. Moreover, a high transverse and depth resolution of around 10 μm may be achieved.

The catheter-endoscope has a unique transverse scanning design and single-mode fibre optics. A lens with a micro-prism is used at the distal end of the catheter-endoscope to emit and accumulate a single spatial-mode optical beam with particular focussing properties. The beam may photograph cross-sections across the structure it is placed into by scanning in a circular pattern. It is possible to create a device as tiny as 1.1 mm in diameter, and images of human venous morphology *in vitro* are described [7].

The layers that make up the normal human oesophagus, which run from the epithelium to the longitudinal muscularis propria, were clearly defined using OCT [8].

A prototype OCT system was created by Colston *et al* to image both soft and hard tissue in the mouth cavity. This method has been used to produce high-resolution pictures of pig periodontal tissues grown *in vitro*. OCT is a potentially helpful method for diagnosing periodontal disorders, as the pictures clearly depict the gingiva-tooth and enamel-cementum interfaces [9]. Using OCT, a quantitative technique for assessing dental caries was developed by Amaechi *et al* in 2001 [10]. It has been demonstrated that stomatological diagnosis is feasible with OCT technology. Due to the polarization and birefringence of teeth, polarization-sensitive OCT in conjunction with polarization technology also retrieves the tooth tissue's polarization features [11].

OCT technology is progressing towards endoscopic imaging with the use of flexible imaging catheters (ICs). The first intravascular OCT system to be sold commercially was created by American Light Lab Imaging Inc. First-generation intravascular OCT (IVOCT) technology obtained a continuous section of blood vessels based on a time-domain OCT system with scanning probe. The Light Lab Imaging Company, USA, and Goodman Company, Japan created the M2-OCT, a second-generation intravascular OCT device, in 2004.

A set of asymmetry measures based on the thickness of the retinal nerve fibre layer (RNFL) in both eyes was proposed by Rafael Berenguer–Vidal [12], who concentrated on primary open-angle glaucoma. The significance of each metric as a pertinent feature for the diagnosis of glaucoma was measured using statistical analysis. They offer an easy-to-use glaucoma screening approach that may be implemented in routine clinical procedures.

Meanwhile in 2000, Fujimoto *et al* described potential biomedical and clinical applications [13]. The integration of OCT with endoscopes and catheters allows for high-resolution intraluminal imaging of organ systems. Because OCT can offer pictures of tissue *in situ* and in real time, unlike conventional histopathology, which requires removing a tissue sample and processing it for microscopic analysis, OCT may be used as a form of optical biopsy and is a potent imaging method for medical diagnostics. OCT can be used to guide non-conventional techniques, minimize sample mistakes associated with excisional biopsy, and replace routine excisional biopsy when it is unsafe or impractical.

A performance analysis of three distinct multiparticulate dose forms with varying architectures (one single-layered, two multi-layered) and layer thicknesses ranging from 5 to 50 μm was conducted by Matthias Wolfgang using ultra-high-resolution (UHR) OCT [14]. The system resolution of 2.4 μm (axial) and 3.4 μm (lateral), both in air, allows for the evaluation of flaws, variations in film thickness, and morphological characteristics in the coating that were not possible to measure with OCT before. The depth of field given was determined to be enough to reach the core region of all dose forms under test, despite the high transverse resolution. The author also showed how to automatically segment and assess UHR-OCT pictures for coating thicknesses, a task that is difficult for human specialists to perform with current OCT devices.

3.2.1 Fingertip biometrics

A non-destructive, high-resolution *in vivo* OCT imaging technique has been modified for the capture of fingertip biometrics. In order to measure information on and beneath the skin, it uses 3D volume data. This data includes the internal fingerprint, sweat pores, and glands, all of which are resilient to counterfeit attempts and bad exterior skin conditions [15].

3.2.2 Urothelial cancer diagnostics

For the diagnosis of urothelial carcinoma in the field of urology, OCT is a promising tool. The possibility of providing equivalent picture quality to traditional OCT systems with miniaturized endoscopic OCT probes was proven by Dominik Stefan Schoeb *et al* [16]. They suggested that a promising method for diagnosing urothelial carcinoma would be to combine various endoscopic imaging modalities.

OCT and Raman spectroscopy on a single device enable biochemical analysis of tissues in addition to structural imaging, providing a more thorough method of clinical diagnosis [17]. In addition to providing insights into key design concerns and data interpretation, the researcher provided an overview of the progress accomplished thus far in integrating spontaneous RS-OCT instruments for biological study.

Mark Draelos *et al* developed a robot-mounted OCT scanner. It is operating safely at a distance to image autonomously and contactlessly without the requirement for operator assistance or head stabilization on the part of independent persons. Optical active scanning is used to find the pupil and attenuate physiological eye movements, while robotic positioning is used to match the scanner with the eye to be scanned. It can be demonstrated that the scanner facilitates the collection of OCT volumetric datasets that resolve important anatomic components that are pertinent to the treatment of common eye disorders. These datasets are on a par with those obtained from clinical tabletop systems. In general practitioners' offices, robotic OCT scanners might make it possible to diagnose and track patients with ocular disorders [18].

A unique imaging method called Spatio-Temporal Optical Coherence Tomography (STOC-T) uses light with regulated spatial and temporal coherence. Without the requirement for mechanical scanning, retinal pictures acquired with the STOC-T technology retain great resolution in all three dimensions on a sample of around 700 μm. By proposing the angio STOC-T approach, Lizewski *et al* improved the depiction of the vasculature in the human eye at different depths by modifying established data processing methods for OCT angiography [19]. The main sensitivity of the algorithms is to the substantial signal phase variance that appears when a large Doppler band appears in STOC-T signals that were acquired with millisecond exposure durations. High contrast pictures of the choroid with the use of STOC-T angiography were recorded.

To help surgeons properly remove tumours from kidneys during surgery, Feng Yan and colleagues devised an optical imaging tool that can recognize tumour zones from normal tissues and clearly determine the borders of kidney tumours.

Polarization-sensitive optical coherence tomography (PS-OCT) was used to scan the kidneys. Kidney sample cross-sectional, enface, and spatial data were acquired for microenvironment reconstruction. H&E histological staining and dice-coefficient were used to measure PS-OCT's efficacy in detecting tumour boundaries and areas in order to confirm the detection accuracy of the method. PS-OCT imaging beat the traditional intensity-based OCT in clearly identifying the tumour margins. Because of the birefringence effects, PS-OCT showed improved contrast of tissue features between normal and malignant tissues. The findings showed that PS-OCT had promise for providing imaging-guided surgical resection guidance for kidney tumours and that it might be used to other human kidney operations in clinics, including renal biopsy [20].

Zhang *et al* presented a dual-modality fibre optic probe that carries out real-time all-optical ultrasound imaging for ablation monitoring in tandem with laser ablation [21]. The apparatus is composed of three optical fibres: one for the supply of laser light for ablation and one for the transmission and reception of ultrasound. Less than 1 mm makes up the entire gadget diameter. Porcine liver and heart tissue were used for *ex vivo* ablation monitoring. A segmentation technique was used to identify the lesion boundaries and all-optical M-mode ultrasound imaging was used to track the ablation depth. Stereomicroscopy was used to visually check the lesions, and results showed good correlation between the ultrasonic depth measures and ablation depths of up to 2.1 mm. The potential of optical ultrasound probes to guide real-time minimally invasive ablation techniques was showcased by the author.

3.3 Optical fibre sensors

Within an optical fibre, light travels by complete internal reflection with comparatively little loss. This makes it possible to transmit data across great distances. The idea of light propagation inside a material by complete internal reflection as a light fountain was originally shown 100 years prior to the first demonstration of a functional fibre optic data transmission system in 1965 [22]. Because of their special qualities, which include its tiny size, ability to detect electromagnetic radiation without interference, and potential for remote sensing, optical fibres have found specialized uses in sensing. Moreover, the technology offers the security required to function in damp or humid situations, like sweating or water-based physical therapy operations, because no current is needed at the place of operating. For continuous monitoring devices in physical applications, these characteristics offer dependable technical support.

To instill a selective response in optical fibre sensors (OFSs), two methods are employed. The initial method relies on evanescent wave (EW) spectroscopy, wherein the interaction between the analyte and EW facilitates the direct examination of the analyte's spectroscopy. However, in order to use this method, specific optical fibres consisting of a material transparent in the infrared wavelength range where biological analytes show significant absorption features are needed. These materials might be silver halide, fluoride, or chalcogenide glasses. Due to the possible toxicity of chalcogenide glasses and the need for more study to fully understand their function in biological systems, the use of these unique optical fibres in the health sector is currently restricted. In the second method, the analyte is measured indirectly by depositing a chemically sensitive functional coating on the surface of the optical fibre, which alters its optical characteristics upon contact with the analyte. It is possible to translate variations in the coating's qualities into variations in the characteristics of the light travelling through the fibre by using the suitable OFS sensor platforms. Quantitative and qualitative data on the chemical species under investigation may then be obtained by analysing the transmission spectrum of the optical fibre.

The sensor's operational wavelength is determined by the characteristics of the chemically sensitive coatings it is used on, as opposed to the absorption spectrum which may be beneficial, the analyte. Because OFSs are based on this method, low-cost light-emitting diodes and photodetectors may be combined to create devices that are inexpensive and small. The sensor area's length determines the device's sensitivity, therefore choosing coating materials with robust optical absorption properties is important for optimal performance.

Chemical substances in the human body can be found in the gas phase by examining gases released via skin, breath, or when expelled. The human body's expelled gaseous components can provide insight into certain metabolic states and blood gas concentrations, potentially leading to the development of non-invasive diagnostic techniques [23]. For the liquid phase, chemical compounds may be found in samples of blood, urine, saliva, perspiration, and tears [23]. The use of colour indicators or fluorophores placed on the tip of an optical fibre is the most commonly used method for pH sensing. The way it works is by detecting changes in the dye indicators' colour caused by pH, the fluorophores' lifespan, or their intensity [24]. The identification of drug concentrations in body fluids would facilitate the proper dosing of pharmacological substances to provide effective therapeutic outcomes in either humans or animals. A promising area of current OFS research is monitoring blood levels of antibiotics [25]. If this approach is successful, it may lead to the possibility of personalized drug treatments, wherein dosages of medication are administered when the blood level of antibiotics approaches the minimal effective dose.

Using a UV-cured pH sensitive hydrogel–[poly(ethylene glycol) diacrylate (PEGDA)] layered on a polymer fibre Bragg grating, Cheng *et al* fabricated a unique optic pH sensor. The volume of the PEGDA expanded in proportion to the pH of the surrounding fluid, causing the polymer fibre Bragg grating to experience a lateral stress. The suggested pH sensor has a quick reaction time of 30 s and a pH sensitivity of up to -0.41 nm pH^{-1} [26]. Maya Chauhan and Vinod Kumar Singh [27] have created a fibre optic pH sensor covered with a sol–gel entrapment indication and a TiO_2–SiO_2 composite acting as a stimuli-responsive layer. The higher sensitivity is attributed to the improved porosity and photoreactivity of a TiO_2–SiO_2 composite film with an optimized volume ratio. The suggested sensor achieves a high sensitivity of 1.76 μW/pH in the pH range of 2–11, together with an increased temperature cross-sensitivity feature. Apart from its exceptional sensitivity, the suggested sensor also boasts swift time response and strong stability, making it highly valuable for usage in industrial, food processing, environmental, and biological settings.

Medical applications have shown a growing interest in wearable optical sensors, a specific subset of OFSs. Fibre optic sensor-based wearables have shown to be efficient and convenient ways to monitor a variety of body functions, including the ankle, blood, spine, heart rate, respiration, neck, shoulder, and joint angle [28].

The classification of wearable OFSs is based on three factors:

1. Fibre Bragg grating (FBG).
2. Light intensity variations.
3. Fabry–Pérot interferometry (FPI).

The FBG sensor consists of a narrow fibre bundle with a lattice pattern evenly carved on its fibre core, cladding layer, and buffer layer. When the core emits broadband light, only particular wavelengths are reflected by the carved lattice pattern. FBG is a hot OFS that can be almost anywhere on a fibre. In the medical industry, FBG is more frequently used to monitor physiological factors such as blood pressure and pulse, temperature monitoring, respiration monitoring systems, and biomechanical studies. In terms of touch perception, FBG sensors do well in measurements. Additionally, their flexibility, small size, non-toxicity, and chemical inertness make them especially well suited for intrusive measurements *in vivo*. Due to its intrinsic magnetic resonance compatibility and tolerance to electromagnetic interference, FBG is the most popular wearable fibre sensor application. In respiratory monitoring, wearable OFSs based on FBGs may achieve an average inaccuracy of 0.33%, however, they are often more costly.

In respiratory monitoring, wearable OFSs with light intensity-based sensors have a little lower accuracy but are less expensive and have a simpler form. Additionally, in order to track finger joint angles for human–machine interaction, they can be included in elastic gloves. Sensors based on variations in optical intensity provide versatility in meeting various joint monitoring needs in the field of joint angle monitoring. This may be accomplished by modifying the light source and receiver's locations and angles to suit different joint configurations.

Wearable OFSs have been used in motion analysis to track joint arches and gait. Fabry–Perot interferometry (FPI)-based wearable OFSs have demonstrated 0.0296 $\pm$ 0.001 mw/° accuracy in gait analysis. These wearables do not restrict the patients' range of motion or interfere with their regular daily activities. Since these physical enablers are in charge of the wearable device's general functionality as well as the data that medical professionals supply for diagnosis, their accuracy must constantly be verified.

A common indirect operating concept is employed by the majority of the biochemical OFSs under investigation. This involves immobilizing a specific recognition element, or receptor, that is responsible for bonding or interacting with target molecules in order to functionalize a particular fibre region. It is vital to discuss optical fibre Raman spectroscopy probes while considering OFSs for *in vivo* application. By measuring a very faint signal from Raman scattering, this label-free method provides nonselective information on the sample's spectral imprint. Even though a number of various Raman spectroscopy OFS probes have been tested *in vivo*, the majority of their applications such as in cancer diagnostics have concentrated on differentiating between normal and diseased tissues. For *in vivo* biosensing, OFS Raman probes continue to be a significant and promising method. Nevertheless, the measurement of Raman scattering is not very selective, which restricts its application to certain parameter measurements. By contrast, the OFSs offer precise and focussed measurements of variables including temperature, pH, and biomolecular analytes.

3.3.1 OFSs for pressure sensor

By using the findings of an intrusive intravascular blood pressure measurement, the fractional flow reserve (FFR) is assessed, which is the most popular *in vivo* application of pressure OFSs in endovascular surgery. In a recent research, vascular surgical procedures in murine models were monitored constantly for vascular pressure using a pressure FPI sensor. In FPI technology, an optical fibre which attaches to a flexible plastic membrane (0.6 mm diameter sensor) coated with gold. This OFS was continually utilized to monitor pressure variations throughout a vascular anastomosis surgical procedure, in contrast to the FFR measurement [29]. It has been demonstrated that the sensor can detect microvascular pressure consistently and in real time for more than an hour. Therefore, in a clinical setting, it might be useful for long-term monitoring of different microsurgical settings.

In animal models, Rodriguez *et al* performed an MRI evaluation of the heart function using an FPI sensor. The sensor was made up of an optical fibre that measured 10 m in length and an FPI that had a silicone membrane positioned on its 0.4 mm diameter and 0.5 mm long tip. A typical angiographic catheter was combined with this OFS, put straight into the sheep's aorta, and pressure was recorded [30].

Since direct intraoperative pressure measurements are challenging, these are often taken in the anterior compartment of the eyeball. A tiny sensor that enables less intrusive implantation into the eyeball might be used to get around this restriction. To achieve intraocular pressure measurement requirements, Cheng *et al* employed a miniaturized FPI-based sensor with a 0.2 mm outside diameter at the distal tip. A tiny 23-gauge needle with an outer diameter of around 0.6 mm was used to implant the sensor directly into the rabbit's eye during microsurgery. Real-time intraocular pressure changes were observed by the sensor, enabling accurate pressure monitoring during the procedure [31].

The majority of pharyngeal manometers in use today are very big, which makes them uncomfortable to use during examinations and may make it difficult to take uninterrupted measurements. Patients will benefit from a more effective and secure measuring process because of the smaller and more flexible pharyngeal manometer made possible by OFS technology. To preserve the assembly, the scientists employed FPI on a distal tip that was encased in a 15 mm long, 0.05 mm thick polytetrafluoroethylene tube that was filled with saline and sealed with a rigid styrene cover. The saline-filled tubing's maximum outer diameter was 2 mm, and its design made it possible to measure pressure from any angle. The sensor was placed just into the patient's throat to track pressures there as they swallowed. Over a broad pressure range, the output of the *in vivo* testing was linearly associated with the reference technique, enabling repeatable findings [32]. In the field of urology, a distal fibre tip-mounted flexible titanium membrane FPI pressure sensor was introduced by Belville *et al* [33]. The sensor was placed into the urine system and incorporated onto the urological catheter sheath. It has undergone *in vivo* testing in a number of clinical settings and been contrasted with a reference technique wherein an adequate resolution of the OFS approach was shown.

In research by Wang *et al*, OFS was used for intravaginal pressure monitoring, demonstrating a novel use for the FBG-based sensor [34]. Two grids of FBG fibres wrapped into a double helix made up the sensor. The FBGs are positioned so that pairs of sensing elements can form an overlap between the two optical fibres. A 20 mm diameter soft material protective sheath encased nine pairs of FBGs, enabling for safe sensor deployment *in situ*. The pairs were arranged in an array at 10-mm intervals. Fibre Optic Real Shape (FORS) technology is a freshly announced strain-sensitive OFS idea that is entirely distinct. To provide the highest level of safety, the sensor is incorporated into the instrument walls so as to avoid making direct contact with the tissues of the patient. With one core running the length of the centre axis and three helical wounds around it, the FORS technology employs numerous FBGs etched along the multicore optical fibre. These pilot trials' results validate the viability and benefit of non-radiation-based OFS technology for endovascular surgical guiding.

3.3.2 Temperature

One of the critical markers that are crucial to understanding our health is our body temperature. In a variety of clinical contexts, such as surgery, cancer therapy, and critical care, it is evaluated utilizing diverse sensing concepts. Especially in high-stress situations where electrical isolation or tolerance to electromagnetic interference is essential, such as during ablation therapy or in MRI settings, OFSs are an excellent substitute for traditional temperature sensors. A few operational principles, including the temperature-dependent fluorescence lifespan, Rayleigh scattering, and the thermo-optical effect and thermal expansion, are used by temperature-sensitive OFSs. *In vivo* brain temperature measurements were conducted using temperature-sensitive OFS based on FBGs. An FBG positioned inside a protective guide cannula and engraved close to the sensor tip made up the OFS. Through a tiny hole created in the skull of an anaesthetised rat model, the sensor which was shielded by a guiding cannula was directly injected into the brain parenchyma. The output of the OFS responded linearly to variations in temperature, according to the experimental data [35]. After acute medication delivery that altered body warmth, it made brain temperature monitoring possible. A new treatment strategy, such as direct monitoring of changes in brain temperature in response to radiosurgery or different drugs, can be made possible by the OFS-based technique, which enables minimally invasive brain temperature monitoring and may provide crucial information on drug activity.

The OFSs and their parameters for physical measures *in vivo* are summarized in table 3.1.

Low-coherence interferometry is used by OCT probes to quantify the optical echoes that are generated when light interacts with the tissue they are imaging. After then, the echoes are processed to provide generic representations of the sample. Despite the fact that OCT OFSs do not offer precise biochemical data.

3.3.3 Other parameters

Recent developments in medicine have raised the need for novel ideas for portable sensors to track patients' vital signs, including blood pressure, respiration, pulse rate, temperature, and more. Since most other sensor technologies are unable to withstand electromagnetic disturbances, equipment that can be used in the very demanding area of strong electromagnetic fields around MRI scanners is particularly interesting. During MRI treatments, for example, a sensor based on optical fibre was used as a 'sandwich' microbend structure to measure breathing and heart rates [36]. Moreover, during sleep monitoring, it was utilized to measure heart activity and breathing. Such a sensor is affixed to the patient's body; and due to mechanical perturbation (external disturbance or movement), the sensitive material (likely an optical fibre) bends, causing a loss of light transmission through the system. The sensor demonstrated high agreement with readings from traditional monitoring devices that were compatible with MRIs. In a laboratory setting, Purnamaningsih *et al* presented a fibre optic macro-bending based respiratory monitoring system for human life activities [37]. Operating at around 1550 nm in wavelength, the respiration sensor is composed of a single-mode optical fibre. An elastic cloth covering the monitored human subject's stomach and chest had an embedded fibre optic. Stretching of the chest and stomach caused deformations in the flexible cloth that included the fibre optic bending curve. A photodetector was utilized to identify the deformation of the fibre, and a PIC18F14K50 microcontroller was employed for processing the data. As a consequence of the walking and running activities, the system was able to exhibit different breathing patterns and rates in real time.

Andersen *et al* presented a luminescence-based OFS [38]. The radiation dosage *in vivo* for patients receiving radiation therapy for cervical cancer was tracked using an array of OFSs. Each probe has an optical fibre of 15 m in length with crystals of aluminium oxide connected. The OFS was inserted using conventional brachytherapy needles into the tumour area. According to the scientists' findings, this sensor configuration could track brachytherapy dosage *in vivo* with a sufficient degree of measurement sensitivity and uncertainty.

Using a modified Michelson interferometer, Djinović *et al* proposed a possibly tiny fibre-optic vibrometer intended to function as a middle-ear microphone for completely implanted cochlear or middle-ear hearing aids. An *in vitro* and *in vivo* study of the acoustical response of sheep middle-ear ossicles was conducted using a model of the sensory system. The use of a middle-ear surgical incision to implant a sensor optical fibre directed towards the incus was studied [39].

Table 3.1. Summary of the OFSs and their parameters for *in vivo* physical measurements. Reprinted from [36], Copyright (2024), with permission from Elsevier.

Parameter	Technology	Purpose	Compartment	Reported results
Pressure	Fabry–Pérot interferometry	FFR (cardiology)	Intravascular	Relative to atmosphere pressure range: –30 to + 300 mmHg; accuracy ±3% of reading or ±3 mmHg
		Blood pressure	Intravascular (microcirculation)	Measured intraluminal mean maximum arterial amplitude: 13.6 mmHg with SD 1.9; mean venous amplitude was 5.2 mmHg with SD 3.3; zero drift during the first 24 h of 0 ± 2 mmHg and less than ±1 mmHg/day on subsequent days
		Intracranial pressure monitoring (neurosurgery)	Intracranial	Maximum zero drift during the first 24 h of 0 ± 2 mmHg and less than ±1 mmHg/day on subsequent days
		Orthopaedics	Intravertebral	Average resting intradiscal pressure: 2.78 ± 0.28 bar, range: 2.31–3.45 bar
		Endoscopy—variceal pressure	Intraoesophegal	Mean variceal pressure value: 21.8 ± 4.4 mmHg.
		Pharyngeal manometry	Intrapharyngeal	Linear correlation with reference method for pressure range 3.0 × 104 to 3.0 × 104 N m^{-2}
	Fibre Bragg grating	Urodynamics	In bladder	Resolution: 100 Pa
		Urodynamics	In bladder	Sensitivity: 1.0–1.6 nm kPa^{-1}; accuracy: 0.3-cm H_2O.
		Bowel manometry	in bowel	Sensitivity: –0.001 nm $mmHg^{-1}$; accuracy: 0.4 kPa (3.1 mmHg)
		Mechanical stress	Knee menisci	Repeatable contact force/stress and pressure measurements in cadaveric joints

		Pressure monitoring	Intravaginal/subbandage	Pressure sensitivity of ∼ 4.8 pm/mmHg.
Strain	Fibre Bragg grating	Needle guidance	Intraparenhymal	Allows for epidural space identification
Shape	Fibre Bragg grating	Intravascular navigation	Intravascular	Allows for determination of 3D conformation of intravascular surgical tools
Temperature	Fibre Bragg grating	Tumour thermal ablation	Intratumoral	Sensitivity ∼3 pm/ °C; measurement error ∼ 3 °C
		Brain temperature	Intraparenhymal	The temperature sensitivity of 12 pm/ °C
	Rare-earth thermometry	Brain temperature	Intraparenhymal	Resolution of 0.1–0.3 °C over the range 23 °C–39 °C.
Breathing rate	Fibre Bragg grating	Patient monitoring (MRI)	Intranasal cannulas	Accuracy ∼ 95% and a relative error ∼ 5%
Respiratory rate	Microbending	Patient monitoring (MRI)	Patient's chest	FOS able to monitor various respiration pattern and rate continuously for a different activity
Multiplexed (breathing + heart activity)	Fibre Bragg grating	Patient monitoring (MRI)	Under patient's back	The relative error span < 7% for respiratory rate; error span of 6.5% (±3 bpm) for cardiac rate
Sound (vibration)	Michelson interferometry	Hearing aids (laryngology)	Inner ear	Amplitude range 10 pm–100 nm
Radiation	Radiation-inducedphoton	absorptionradioluminescence	Therapeutic irradiation	Orbit (eye)
		Scattered doses of 2.7 ± 0.5 cGy		
Brachytherapy	Prostate	A maximum measured dose departure of 9% from the calculated dose		

Wittauer *et al* created a portable fibre-optic instrument for intramuscular oxygen measurements [40]. The apparatus relies on phosphorescence quenching, in which an optical fibre's tip is coated with a poly(propyl methacrylate) (PPMA) matrix that contains a Pt(II)-core porphyrin that emits light brilliantly. This very portable optoelectronic circuit uses a microspectrometer and a microcontroller readout that can be accessed through a smartphone. The sensor is sensitive across the physiological oxygen partial pressure range of 0–80 mmHg, according to data from an *in vivo* tourniquet porcine model, and it responds appropriately and consistently to variations in intramuscular oxygen. An optical fibre tip is connected to a ruthenium luminophore via a silicone rubber polymer, as per the sensor proposed by Seddon *et al* [41]. After the ruthenium luminescence was excited by a blue light-emitting diode, a photomultiplier tube was utilized to detect it. Once more, that study showed how useful OFS is for evaluating low-oxygen situations. It is noteworthy that the biocompatibility of the sensor takes a backseat in the event that it is inserted into malignant neoplastic tissue. The possibility of the neoplastic process spreading during the sensor implantation must, however, be taken into account.

Fluorescence dye-based OFSs are one type of sensor that can track changes in blood pH over time. Usually, weak electrolyte dyes that are pH-sensitive are used in such pH-sensitive OFSs. The relative amounts of dyes in basic and acid forms can be expressed using PH dependent dissociation curves. Since the amounts of the dye present in the protonated and deprotonated forms vary with the quantity of H^+ ions, variations in fluorescence strength are consequently connected with the pH of a sample. A sensor based on a proton-sensitive fluorescent dye bonded to the optical fibre tip by heat polymerization was presented by Jin *et al* [42].

Rajash *et al* reported measuring the pH of a big mammal's intraluminal papillary duct urine *in vivo* using an optic-chemo microsensor [43]. With a tip diameter of 140 μm, fibre-optic pH microsensors may be inserted into papillary Bellini ducts to monitor the proton content of urine in tubules. Adult pigs under anaesthesia had a percutaneous nephrolithotomy to reach the urine collecting system's lower pole. The working channel of a flexible nephroscope containing a fibre-optic microsensor was moved towards an upper pole papilla. The pH of the tubule urine was then continuously measured by inserting the microsensor, very gently, into the Bellini ducts. The pH levels of tubular urine were effectively measured in five papillary ducts from three pigs—two metabolic syndrome Ossabaw pigs and one farm pig. Their findings show that intraluminal urine pH may be measured in real time in a live big animal using optical microsensor technology. This makes the use of this optical pH sensor method in nephrolithiasis possible.

One fascinating area of continuing OFS research that aims to enable personalized pharmacological therapy is the assessment of drug concentration in patients' blood. Sensors that are able to detect the concentration of medications of interest in an *in vivo* situation selectively are in great demand. The OFS tip was exposed and protected with a polyporous membrane that could release fluorescence. Adriamycin in the blood subsequently quenched the fluorescence, enabling real-time drug concentration measurements [44]. The probe exhibited a detection limit as low as 0.057 $\mu g\ ml^{-1}$. When measuring the concentration of adriamycin, it demonstrated good agreement with the reference method. An OFS of this kind has a significant advantage over traditional detection techniques: it can track the concentration of the target molecule in real time and doesn't require sample preparation.

In the rapidly developing field of artificial pancreas concepts, where continuous monitoring of blood glucose levels is required, glucose detecting probes are essential. Many wearable sensors that detect interstitial glucose levels constantly, provide guidance for decision-making, and improve the treatment of diabetic patients have been commercially accessible in recent years [45]. The majority of continuous glucose monitoring devices employs an electrochemical method. Nonetheless, OFS is a compelling substitute medium that has noteworthy advantages, including resistance to electromagnetic interference and the potential for multiplexing. In order to detect different glucose concentrations, tapered optical fibres as refractive index sensors have been developed and proven by Ujah *et al* [46]. An optical fibre can be made tapered by slowly stretching it while it is heated, like over a flame, until the glass softens. Over a certain distance, such as a few millimetres or centimetres, this process thins the fibre. The same element that thins the fibre as a whole also thins the fibre core. As the diameter of the taper waist decreases and the difference between the RI of the fibre and the surrounding medium decreases, so does the proportion of power in the evanescent field and the interaction with the medium. The sensor's power output intensity dropped as the glucose concentration rose. The sensor's sensitivity went enhanced by around four times when AuNPs were added to the tapered optical fibre ($\varnothing = 12$ μm), however the linear RI range marginally shrank. The authors proposed robust, repeatable, and straightforward fabrication method [45]. A difunctional fluorescent hydrogel optical fibre was created by Li *et al* [47] to monitor pH and glucose continuously at the same time. Quantum dots and a derivative of fluorescein were used to segmentally functionalize the hydrogel fibre using 3-(acrylamido) phenylboronic acid (3-APBA). Quantum dots/3-APBA and the sensitive agent fluorescein derivative stand for glucose and pH detection, respectively. By connecting 490 nm light into the hydrogel fibre via the multimode fibre, two fluorophores were stimulated, and the fluorescence was returned to the spectrometer. Moreover, the sensor can detect glucose and pH simultaneously thanks to its two distinct emissions, which are 594 nm of QDs and 517 nm of a fluorescein derivative. By measuring the intensity of the fluorescence, we can determine the pH and glucose levels. In contrast, pH monitoring is a calibration used to increase the accuracy of glucose detection. The sensor can constantly measure pH between 5.4 and 7.8 and glucose between 0 and 20 mM. Besides, the difunctional fluorescent hydrogel optical fibre may be employed as a dynamic sensor because of the reversibility of the two sensing processes. The experiment including continuous, simultaneous monitoring shows that the sensor is very tolerant to fluctuations in pH and glucose content.

3.4 Future perspectives and challenges

Owing to its special qualities, optical fibre technology might revolutionize healthcare by enabling less invasive *in vivo* insertions. Optical fibre might be used to monitor certain parameters in real time in difficult-to-reach locations, including the lumen of aneurysms or the microcirculation of tumours, which are inaccessible to the diagnostic instruments now in use. Such first-hand evaluations of the local micro-environment would produce unique data to follow treatment success, develop new prognostic algorithms, and learn more about the pathophysiology of the disease. Thus far, single-target detection has been the main emphasis of most sensing approaches. Since distinct disease indicators signify various illnesses, multiplexed detection of numerous biomarkers is an attractive option. Although multiplexed OFS assay development is still in its infancy, further studies are anticipated to further this field. This method falls within the category of 'lab-on-a-fibre' sensing approaches, which are the subject of much current research. This technology's main concept is the switch from a basic optical fibre to a multifunctional detection system made up of an integrated structure that offers several target functionality. The OFS shows great promise for advancing the 'lab-on-a-fibre' idea because of its multi-plexing capabilities. Lastly, the materials used in the present OFS designs are non-resorbable and permanent, and thus need to be removed after the measurements. The creation of the upcoming generation of biocompatible sensor components is the result of recent advances in materials engineering and optoelectronics. There may be new avenues for the advancement of optical fibre technology now that the first viable bioresorbable OFS prototypes have been released.

References

[1] Tosi D, Poeggel S, Iordachita I and Schena E 2018 Fiber optic sensors for biomedical applications *Opto-mechanical Fiber Optic Sensors: Research, Technology, and Applications in Mechanical Sensing* (Butterworth-Heinemann) pp 301–33

[2] Fercher A F, Hitzenberger C K, Drexler W, Kamp G and Sattmann H 1993 *In vivo* optical coherence tomography *Am. J. Ophthalmol.* **116** 113–4

[3] Swanson E A, Izatt J A, Lin C P, Fujimoto J G, Schuman J S, Hee M R, Huang D and Puliafito C A 1993 *In vivo* retinal imaging by optical coherence tomography *Opt. Lett.* **18** 1864

[4] Izatt J A, Hee M R, Swanson E A, Lin C P, Huang D, Schuman J S, Puliafito C A and Fujimoto J G 1994 Micrometer-scale resolution imaging of the anterior eye *in vivo* with optical coherence tomography *Arch. Ophthalmol.* **112** 1584–9

[5] Colston B W, Sathyam U S, DaSilva L B, Everett M J, Stroeve P and Otis L L 1998 Dental OCT *Opt. Express* **3** 230

[6] Baumgartner A, Dichtl S, Hitzenberger C K, Sattmann H, Robl B, Moritz A, Fercher A F and Sperr W 2000 Polarization-sensitive optical coherence tomography of dental structures *Caries Res.* **34** 59–69

[7] Tearney G J, Boppart S A, Bouma B E, Brezinski M E, Weissman N J, Southern J F and Fujimoto J G 1996 Scanning single-mode fiber optic catheter-endoscope for optical coherence tomography *Opt. Lett.* **21** 543–5

[8] Bouma B E, Tearney G J, Compton C C and Nishioka N S 2000 High-resolution imaging of the human esophagus and stomach *in vivo* using optical coherence tomography *Gastrointest. Endosc.* **51** 467–74

[9] Colston B W, Everett M J, Da Silva L B, Otis L L, Stroeve P and Nathel H 1998 Imaging of hard- and soft-tissue structure in the oral cavity by optical coherence tomography *Appl. Opt.* **37** 3582

[10] Amaechi B T, Higham S M, Podoleanu A G, Rogers J A and Jackson D A 2001 Use of optical coherence tomography for assessment of dental caries: quantitative procedure *J. Oral Rehabil.* **28** 1092–3

[11] Fried D, Xie J, Shafi S, Featherstone J D B, Breunig T M and Le C 2002 Imaging caries lesions and lesion progression with polarization sensitive optical coherence tomography *J. Biomed. Opt.* **7** 618

[12] Berenguer-Vidal R, Verdú-Monedero R, Morales-Sánchez J, Sellés-Navarro I, Kovalyk O and Sancho-Gómez J L 2022 Decision trees for glaucoma screening based on the asymmetry of the retinal nerve fiber layer in optical coherence tomography *Sensors (Basel)* **22** 4842

[13] Fujimoto J G, Pitris C, Boppart S A and Brezinski M E 2000 Optical coherence tomography: an emerging technology for biomedical imaging and optical biopsy *Neoplasia* **2** 9–25

[14] Wolfgang M, Kern A, Deng S, Stranzinger S, Liu M, Drexler W, Leitgeb R and Haindl R 2023 Ultra-high-resolution optical coherence tomography for the investigation of thin multilayered pharmaceutical coatings *Int. J. Pharm.* **643** 123096

[15] Yu Y, Wang H, Sun H, Zhang Y, Chen P and Liang R 2022 Optical coherence tomography in fingertip biometrics *Opt. Lasers Eng.* **151** 106868

[16] Schoeb D S, Wollensak C, Kretschmer S, González-Cerdas G, Ataman C, Kayser G, Dressler F F, Gratzke C, Zappe H and Miernik A 2022 Ex-vivo evaluation of miniaturized probes for endoscopic optical coherence tomography in urothelial cancer diagnostics *Ann. Med. Surg.* **77** 103597

[17] Fitzgerald S, Akhtar J, Schartner E, Ebendorff-Heidepriem H, Mahadevan-Jansen A and Li J 2023 Multimodal Raman spectroscopy and optical coherence tomography for biomedical analysis *J. Biophotonics* **16** e202200231

[18] Draelos M, Ortiz P, Qian R, Viehland C, McNabb R, Hauser K, Kuo A N and Izatt J A 2021 Contactless optical coherence tomography of the eyes of freestanding individuals with a robotic scanner *Nat. Biomed. Eng.* **5** 726–36

[19] Liżewski K, Tomczewski S, Borycki D, Węgrzyn P and Wojtkowski M 2024 Imaging the retinal and choroidal vasculature using spatio-temporal optical coherence tomography (STOC-T) *Biocybern. Biomed. Eng.* **44** 95–104

[20] Yan F, Wang C, Yan Y, Zhang Q, Yu Z, Patel S G, Fung K M and Tang Q 2024 Polarization-sensitive optical coherence tomography for renal tumor detection in *ex vivo* human kidneys *Opt. Lasers Eng.* **173** 107900

[21] Zhang S, Zhang E Z, Beard P C, Desjardins A E and Colchester R J 2022 Dual-modality fibre optic probe for simultaneous ablation and ultrasound imaging *Commun. Eng.* **1** 9

[22] Hecht J 1999 *City of Light: The Story of Fiber Optics (Sloan Technology Series)* (Oxford: Oxford University Press)

[23] Correia R, James S, Lee S W, Morgan S P and Korposh S 2018 Biomedical application of optical fibre sensors *J. Opt.* **20** 073003

[24] Wolthuis R, McCrae D, Saaski E, Haiti J and Mitchell G 1992 Development of a medical fiber-optic pH sensor based on optical absorption *IEEE Trans. Biomed. Eng.* **39** 531–7

[25] Korposh S, Chianella I, Guerreiro A, Caygill S, Piletsky S, James S W and Tatam R P 2014 Selective vancomycin detection using optical fibre long period gratings functionalised with molecularly imprinted polymer nanoparticles *Analyst* **139** 2229–36

[26] Bonefacino J, Tam H Y, Cheng X and Guan B O 2018 All-polymer fiber-optic pH sensor *Opt. Express* **26** 14610–6

[27] Chauhan M and Singh V K 2020 Fiber optic pH sensor using TiO_2–SiO_2 composite layer with a temperature cross-sensitivity feature *Optik (Stuttg)* **212** 164709

[28] Zhang X, Wang C, Zheng T, Wu H, Wu Q and Wang Y 2023 Wearable optical fiber sensors in medical monitoring applications: a review *Sensors 2023* **23** 6671

[29] Walle L, Sudhoff H, Frerichs O and Todt I 2021 Intraluminal monitoring of micro vessels. A surgical feasibility study *Front. Surg.* **8** 681797

[30] Abi-Abdallah Rodriguez D, Durand E, de Rochefort L, Boudjemline Y and Mousseaux E 2015 Simultaneous pressure-volume measurements using optical sensors and MRI for left ventricle function assessment during animal experiment *Med. Eng. Phys.* **37** 100–8

[31] Cheng W *et al* 2018 Real-time intraocular pressure measurements in the vitreous chamber of rabbit eyes during small incision lenticule extraction (SMILE) *Curr. Eye Res.* **43** 1260–6

[32] Takeuchi S, Tohara H, Kudo H, Otsuka K, Saito H, Uematsu H and Mitsubayashi K 2007 An optic pharyngeal manometric sensor for deglutition analysis *Biomed. Microdevices* **9** 893–9

[33] Belville W D, Swierzewski S J, Wedemeyer G, McGuire E J *et al* 1993 Fiber optic microtransducer pressure technology: Urodynamic implications *Neurourol. Urodyn.* **12** 171–8

[34] Wang D H-C *et al* 2013 An optical fiber Bragg grating force sensor for monitoring sub-bandage pressure during compression therapy *Opt. Express* **21** 19799–807

[35] Zibaii M I, Latifi H, Karami F, Ronaghi A, Nejad S C and Dargahi L 2017 *In vivo* brain temperature measurements based on fiber optic Bragg grating *25th Int. Conf. on Optical Fiber Sensors* (Bellingham, WA: SPIE) pp 639–42

[36] Bartnik K, Koba M and Śmietana M 2024 Advancements in optical fiber sensors for *in vivo* applications—a review of sensors tested on living organisms *Measurement* **224** 113818

[37] Purnamaningsih R W, Widyakinanti A, Dhia A, Gumelar M R, Widianto A, Randy M and Soedibyo H 2018 Respiratory monitoring system based on fiber optic macro bending *AIP Conf. Proc.* **1933** 40012

[38] Andersen C E, Nielsen S K, Lindegaard J C and Tanderup K 2009 Time-resolved *in vivo* luminescence dosimetry for online error detection in pulsed dose-rate brachytherapy *Med. Phys.* **36** 5033–43

[39] Djinović Z, Pavelka R, Tomić M, Sprinzl G, Plenk H, Losert U, Bergmeister H and Plasenzotti R 2018 In-vitro and in-vivo measurement of the animal's middle ear acoustical response by partially implantable fiber-optic sensing system *Biosens. Bioelectron.* **103** 176–81

[40] Witthauer L, Cascales J P, Roussakis E, Li X, Goss A, Chen Y and Evans C L 2021 Portable oxygen-sensing device for the improved assessment of compartment syndrome and other hypoxia-related conditions *ACS Sens.* **6** 43–53

[41] Seddon B M, Honess D J, Vojnovic B, Tozer G M and Workman P 2001 Measurement of tumor oxygenation: *in vivo* comparison of a luminescence fiber-optic sensor and a polarographic electrode in the P22 tumor *Radiat. Res.* **155** 837–46

[42] Jin W, Wu L, Song Y, Jiang J, Zhu X, Yang D and Bai C 2011 Continuous intra-arterial blood pH monitoring by a fiber-optic fluorosensor *IEEE Trans. Biomed. Eng.* **58** 1232–8

[43] Handa R K, Lingeman J E, Bledsoe S B, Evan A P, Connors B A and Johnson C D 2016 Intraluminal measurement of papillary duct urine pH, *in vivo*: a pilot study in the swine kidney *Urolithiasis* **44** 211–7

[44] xu Lu W and Chen J 2003 Continuous monitoring of adriamycin *in vivo* using fiber optic-based fluorescence chemical sensor *Anal. Chem.* **75** 1458–62

[45] Yu Z, Jiang N, Kazarian S G, Tasoglu S and Yetisen A K 2021 Optical sensors for continuous glucose monitoring *Prog. Biomed. Eng.* **3** 022004

[46] Ujah E, Lai M and Slaughter G 2023 Ultrasensitive tapered optical fiber refractive index glucose sensor *Sci. Rep.* **13** 4495

[47] Li Y, Luo S, Gui Y, Wang X, Tian Z and Yu H 2023 Difunctional hydrogel optical fiber fluorescence sensor for continuous and simultaneous monitoring of glucose and pH *Biosensors* **13** 287

Chapter 4

Revolutionizing dentistry: applications of bioactive glass

Manoj Kumar Srinivasan, Saravanan Alamelu, Briska Jifrina Premnath, Nivedha Jayaseelan, Kamalesh Balakumar Venkatesan, Sathish-Kumar Kamaraj and Kalist Shagirtha

As research progresses, there have been notable advancements in the realm of dental materials, encompassing both the creation of novel materials and the enhancement of existing materials to boost their performance. Through this process, natural ground substance, tissue components, and durable tissues undergo growth. A prominent biomaterial currently in use is referred to as bioactive glass (BAG). The inherent bioactive characteristics of BAG make it suitable for various clinical applications, particularly in the field of medicine and dentistry, where it aids in the regeneration of hard tissues. BAG preparation methods include melt-quenching and sol–gel techniques, each with its advantages and limitations. BAG facilitates the formation of hydroxyapatite (HA), establishing connections with hard tissues. They can be osteoconductive or osteoinductive, promoting bone growth or stimulating bone tissue formation. Enhanced bioactivity is associated with larger surface area, higher dissolution rates, and quicker apatite formation. However, limitations in mechanical properties and cytotoxicity of some BAG need addressing. In dentistry, BAGs are used for dental implants, dentin hypersensitivity treatment, and as bone grafts to promote osteogenesis. This chapter underscores the multifaceted potential of BAG in dentistry, while also highlighting the future perspectives of BAG.

4.1 Introduction

BAGs are a type of non-crystalline ceramics known for their unique ability to bond with living tissues, facilitate the formation of new tissues, and gradually degrade over time. These remarkable properties make them promising candidates for various tissue engineering applications. Initially developed to address bone defects, BAGs have expanded their biomedical versatility, particularly through specific

doi:10.1088/978-0-7503-5904-7ch4

compositions like Bioglass® 45S5, which exhibits bonding capabilities with both hard and soft tissues [1].

The convergence of mechanical biocompatibility and physiological activity, exemplified by the inclusion of inorganic HA, known for its affinity with calcified tissues [2], underscores the profound significance of BAG in research and application. For instance, these materials find extensive utility in dentistry, where they serve as coatings for implants, bone grafts, restorative materials, and scaffolds for tissue engineering [3–10].

Biomaterials, whether natural or synthetic, are substances designed to interact with biological systems and find use in implant construction and tissue regeneration. In the realm of implant biomaterials, a range of metallic materials, including stainless steel 316L, cobalt–chromium alloys, titanium (Ti), and Ti–6Al–4V alloy, are employed [11]. These biomaterials aim to enhance the quality of life for patients by replacing or restoring damaged or deteriorated tissues and organs [12].

Within the emerging field of nanomedicine, nanotechnology is making significant strides in addressing various medical engineering and biomedical challenges. The nanoscale dimensions inherent to biological tissues, along with the critical role of nanoscale interactions between cells and biomaterial surfaces (e.g., at 100 nm scale), render nanoscience and nanotechnology particularly appealing in the context of regenerative medicine and tissue engineering approaches [13, 14].

Over the past four decades, BAGs have garnered attention due to their exceptional osteoconductivity, the ability to promote bone growth on implant surfaces, and their customizable degradation rates. These attributes have led to their consideration for use in orthopedic devices, dental procedures, and increasingly, in bone tissue engineering applications [15, 16].

4.2 Bioactive glass compositions

BAGs, were first developed in 1969, with the initial hypothesis that they would degrade in bodily fluids, yielding hydroxycarbonated apatite (HCA) [17]. While originally designed for bone substitution, they have gained popularity in the past fifteen years, particularly as an ingredient in toothpaste formulations aimed at remineralizing and desensitizing hypersensitive teeth [18–21]. These glass compositions are primarily rooted in the SiO_2–P_2O_5–CaO–Na_2O system. Table 4.1 provides some of the typical BAG compositions.

The dissolution and apatite formation mechanism, as extensively elucidated by Hench *et al* (figure 4.1) [17]. However, it may be concluded from this procedure that even window glass, a Q^3 or Q^4 type glass distinguished by three to four bridging oxygens bonded to silicon, can display bioactivity. It was eventually discovered, however, that glasses often start to become bioactive at around two network connections (NCs), which corresponds to Q^2 type glass, where there are two bridging oxygens bound to silicon. Consequently, bioactivity is greatly influenced by NC concentration. It is crucial to keep in mind that this mechanism only takes into account Na^+ cations, although in actuality, additional cations, most notably Ca^{2+}, are also present. Although different apatite phases, such as fluorapatite

Table 4.1. Some important BAG compositions.

Glass	Mol%					NC	RI	References
	CaO	Na_2O	CaF_2	SiO_2	P_2O_5			
BAG-F	26.20	26.10	1.00	42.70	4.00	2.11	1.55	[22]
45S5F	16.90	24.40	10.00	46.10	2.60	2.55	1.53	[23]
45S5	26.90	24.40	0.00	46.10	2.60	2.11	1.56	[17]
S53P4	21.80	22.70	0.00	53.80	1.70	2.54	1.54	[24]
QMNA1	43.00	6.00	10.00	35.00	6.06	2.24	1.57	[25]
BioMin-F	28.00–30.00	22.00–24.00	1.50–3.00	36.00–40.00	4.00–6.00	2.16	1.52	[26]
Example 7 (Cention N)	31.00	8.00	10.00	48.00	0.00	2.38	1.50	[27]

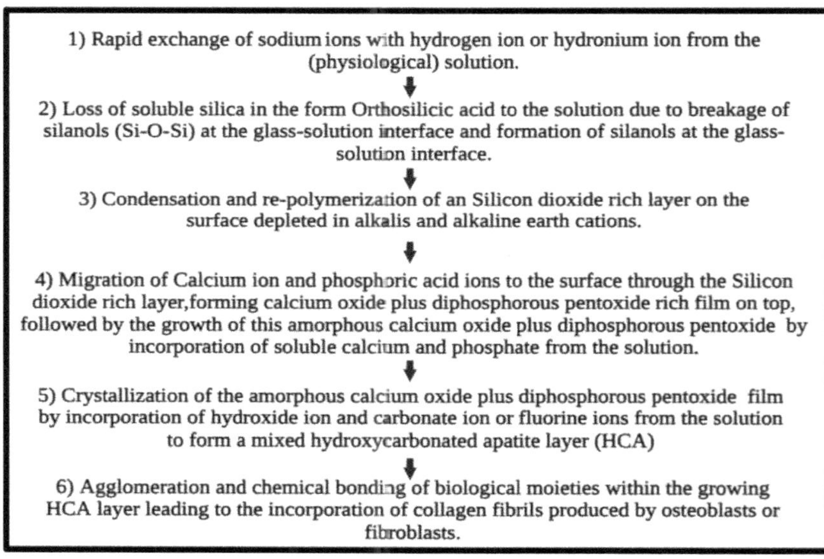

Figure 4.1. Mechanism of dissolution and apatite formation of BAG.

(FAp), chlorapatite (HAp), strontium-substituted apatite (SSAT), and mixed apatites, can form, this mechanism primarily focuses on HCA formation and does not take into account the incorporation of minute amounts of CaF_2 into the glass composition [28].

Notably, due to the influence of OH-ions on the lattice structure, FAp ($Ca_5(PO_4)_3F$) is more resistant to acidity and less susceptible to degradation compared to both HCA and Hap [29]. In contrast, F-ions, with their smaller ionic radius, fit snugly within the hexagonal apatite lattice. Given the frequent exposure of teeth to acidic conditions in the mouth, fluoride becomes particularly interesting in the field of dentistry [30]. For individuals, the critical pH values for FAp and HAp are approximately 4.5 and 5.5, respectively [31].

4.3 Bioactive glass preparation methods

Traditionally, BAG like Bioglass® 45S5 have been crafted using a melt-quenching technique [32]. This process entails heating powdered substances to temperatures exceeding 1300 °C, followed by rapid cooling to freeze the atomic structure. However, this method presents some challenges, such as the difficulty in creating porous scaffolds and reduced bioactivity at higher sintering temperatures [33]. Heat treatment is frequently used on the glass to reduce thermomechanical strains brought on by rapid cooling. However as demonstrated by the thermal treatment of Bioglass® 45S5, heat treatment can, at certain temperature ranges, cause the creation of different crystalline phases that may adversely affect elastic modulus and strength, increasing the likelihood of mechanical failure. To alleviate such concerns, silicate-based BAG is subjected to heat treatment to release internal stresses from the glass [34]. It is worth noting that when glass particles are sintered into glass-ceramic scaffolds, this can limit bioactivity and ion dissolution due to crystallization.

Beginning in the early 1970s, a different technology for creating BAG called the sol–gel technique was developed [35]. With this method, a variety of glass compositions and structures, such as fibers, coatings, scaffolds, and nanoparticles, can be produced [36]. In one noteworthy application, Midha *et al* created BAG scaffolds (consisting of 70% SiO_2 and 30% CaO, designated as 70S30C) using a sol–gel foaming process that shows potential as a matrix for bone tissue regeneration (figure 4.2) [37]. Precursors are used in this process, which includes a sequence of condensation and hydrolysis processes as well as low-temperature heat treatments. Sol–gel glasses have more porosity, an improved capacity to promote apatite formation, and a bigger surface area in comparison to melt-quenched glasses, which boast superior mechanical qualities [38].

4.4 Characteristics of bioactive glass

Remarkably, BAG can interact with its biological surroundings and trigger particular biological reactions. For instance, it can promote the development of a HA layer, creating a crucial link between the substance and the tissue around it. These interactions predominantly occur within hard tissues, such as enamel, dentin in bones and teeth, and HA, a crystalline calcium phosphate [39]. In contrast, bioinert materials remain inert within the biological milieu, eliciting no particular responses. However, they may trigger foreign-body reactions, leading to the formation of fibrous capsules. Over time, these fibrous capsules can lead to the failure of prostheses due to micro-movements they induce. Bioactive compounds can fall into two categories: osteoconductive, which promote bone growth, and osteoinductive, which stimulate the formation of bone tissue [16].

The degree of bioactivity in glass is influenced by several factors. Glass with enhanced bioactivity typically possesses a higher dissolution rate, larger surface area, and a quicker formation of apatite [40]. Furthermore, these BAGs exhibit

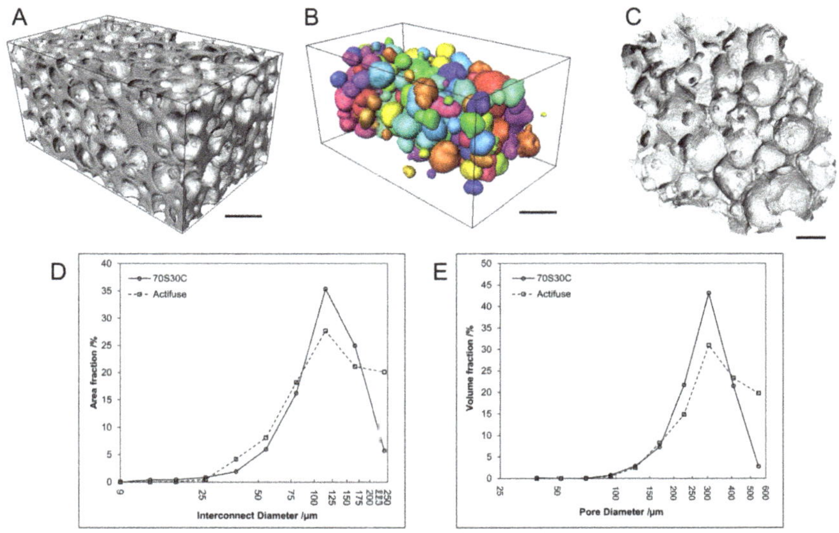

Figure 4.2. Material characterization employing micro-CT showing (A) reconstruction of a child volume of the 70S30C foam; (B) the separated pores from a child volume of a piece of 70S30C; (C) distribution of interconnect sizes displayed as area fraction (for interconnects); (D) distribution of interconnect sizes displayed as area fraction. (E) Interconnect and pore size distributions of 70S30C and Actifuse determined from micro-CT images. Scale bar = 400 μm. Reprinted from [37] CC BY 3.0.

improved mechanical properties when compared to natural bone, providing biomimetic nanostructures that enhance cell adhesion.

The bioactive qualities of glass are also contingent upon its structural composition, manufacturing processes, and the rate at which it undergoes ionic dissolution. A notable comparison can be made between BAG and the conventional Bioglass® 45S5, which highlights the limitations of the latter. Bioglass® 45S5 is characterized by its rapid degradation, which can create gaps between the material and host tissues, among other drawbacks [41, 42]. Apart from composition, the method of application and the level of particle aggregation also play pivotal roles in determining the presence or absence of porosity [43, 44]. Additionally, Bioglass® 45S5 can exhibit cytotoxic effects due to the high pH levels resulting from the leakage of Na^+ and $Ca2^+$. This, in turn, can lead to a delayed synthesis of HA [45]. Furthermore, because of the intrinsic fragility of glass, its mechanical qualities may not be sufficient for the creation of porous scaffolds [46, 47]. To harness the full potential of BAG, further research is imperative to enhance its mechanical characteristics and address the mentioned limitations.

BAG possesses remarkable antimicrobial properties due to its ability to elevate the pH of aqueous solutions through cation release, creating an inhospitable environment for most microorganisms [48]. A study conducted by Allan *et al* delved into the role of BAG as an effective antibacterial agent in treating periodontal abnormalities. Their findings revealed that BAG, by providing a source of calcium ions and elevating the local pH, effectively inhibits bacterial colonization [49].

Further substantiating its antimicrobial prowess, S53P4 forms of BAG have demonstrated the capability to eliminate *Streptococcus mutans*, *Actinomyces naeslundii*, and *Aggregatibacter actinomycetemcomitans*. These bacteria are associated with enamel caries, root caries, and periodontitis, respectively, underscoring the potential of BAG in combatting oral infections [50]. Notably, inadequate control of bacterial counts remains a common factor contributing to failures in endodontic and periodontal treatments. Thus, the significance of antibacterial and disinfection agents in clinical dentistry cannot be overstated [51, 52].

4.5 Clinical applications of BAG in dentistry

BAGs offer a multitude of clinical applications within the realms of both medicine and dentistry. In the dental domain, they are commonly employed as a coating material for dental implants and as an ingredient in dentifrices to address dentin hypersensitivity (DH). Meanwhile, in the field of medicine, they are often utilized as a bone graft to promote the formation of new bone tissue through osteogenesis [4, 45, 53]. Here, we'll delve into several typical applications of BAG in dentistry (figure 4.3).

4.5.1 Bone grafts

When bone loss occurs due to infection, trauma, or disease, bone grafts are employed for replacement [54]. Bone grafting has been employed as a material for bone transplantation for over two decades. It stimulates the growth of new bone on its surface and exhibits superior osteoconductive characteristics [55]. A prior investigation compared HA and bone graft (BG) as bone graft materials in an animal model, revealing that BG not only offers ease of handling but also accelerates bone restoration within two weeks, whereas HA necessitates 12 weeks for a comparable outcome [56]. A comprehensive review of the literature, which included

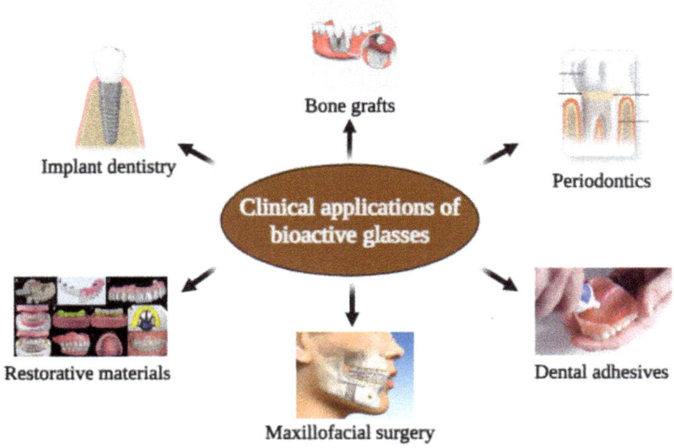

Figure 4.3. Clinical applications of bioactive glass in dentistry.

data from several long-term follow-up studies, similarly confirmed the exceptional bone-healing properties of BG when used as a bone transplant material [57].

4.5.2 Bone regeneration

The prevalence of age-related bone abnormalities is expected to increase in line with the growing aging population, leading to a higher demand for artificial bone graft substitutes. Various factors such as trauma, congenital or developmental issues, deformities, malignancies, post-operative complications, periodontitis, or osteomyelitis can lead to bone defects [58, 59]. These skeletal anomalies may cause economic hardships and a lower quality of life.

Around 700 000 of the roughly 2 million bone transplant surgeries carried out annually throughout the world are used to mend the cranial bones [59]. At the moment, autologous bone grafts, allogenic BAG, xenografts, and synthetic BAG are the materials used for bone grafting. Due to their capacity to offer all the components required for efficient bone regeneration, their low risk of negative immunological reactions, and their great osteogenic potential, autologous BAGs are regarded as the gold standard [58, 60]. However, their use is limited by the quantity of graft that can be harvested, which can lead to a secondary bone defect and donor site morbidity [61, 62].

In the context of the criteria set, neither autologous, allogenic BAG, nor xenografts meet the standards of an ideal grafting material [60]. Synthetic BAGs are gaining popularity due to the increasing demand, offering an advantage because they can be custom-designed to possess optimal characteristics as a bone-grafting material [63].

4.5.3 Implant dentistry

In the field of implant dentistry, endosseous implants, commonly referred to as dental implants (DIs), are cylindrical devices inserted into the alveolar bone to support dental prosthetic structures, thereby enhancing both functionality and esthetics [64]. Osseointegration, which requires direct contact between the implant's surface and the bone tissue to guarantee adequate anchorage inside the bone structure, is a crucial requirement for its success [65]. The preferred material for manufacturing DIs is titanium-based alloys. Despite their bioinert nature, these alloys exhibit remarkable osteoconductive properties and excellent biocompatibility [45]. They serve as a platform for osteoblasts, which are crucial osteoprogenitor cells, while simultaneously being toxic to microorganisms [66, 67]. However, there have been instances of titanium-DIs failing to achieve osseointegration [68, 69]. In order to enable successful osseointegration, DIs must be sufficiently strong mechanically to withstand the stresses generated while chewing, not trigger inflammatory or foreign body reactions, and actively encourage bone attachment. To maintain stability and osseointegration and prevent early implant failure, a healing time of three to six months is often needed [70, 71]. The integration of BAG, which has the ability to promote active bonding between bone and implant, provide antimicrobial protection, and reduce the overall treatment time, may benefit Ti-DIs' intrinsic bioinertness [67, 72, 73]. It is important to note that there aren't any therapeutic BAG coatings for DIs on the market right now.

One significant challenge in applying BAG coatings to DIs lies in the difference in thermal expansion coefficients (TECs) between BAG and metal. This disparity can lead to cracking in the coating due to differential shrinkage during cooling. In order to prevent these problems and prevent breaking, the glass should ideally have a slightly lower TEC than the metal [74]. Silica content can be increased, or CaO and Na_2O can be partially replaced with MgO and K_2O to adjust the glass TEC [52, 75].

Various surface deposition techniques have been explored to develop reliable BAG coatings for DIs, including glazing [75–78], sol–gel deposition [79, 80], electrophoretic deposition [81, 82], pulsed laser deposition [83, 84], ion-beam methods [85], and radio-frequency magnetron sputtering (RF-MS) [86–88]. Among these techniques, RF-MS has demonstrated promising results, producing coatings characterized by exceptional adhesion and purity, even on complex geometric implant structures [89, 90]. *In vivo* animal tests have shown that BAG-coated Ti-DIs osseointegrate significantly better with the surrounding bone tissue compared to control DIs [91, 92]. In particular, deceased pig mandibular bones exhibiting uniform and mechanically resilient BAG coatings, created using RF-MS biocompatibility experiments, demonstrated strong cellular adherence and proliferation of dental pulp stem cells (dDPSCs), highlighting the pivotal role played by BAG in advancing the next generation of DIs [93].

4.5.4 Maxillofacial surgery

There has been a lot of study done on the use of Bioglass® 45S5 and comparable formulations in the field of maxillofacial surgery. Compared to other calcium phosphate-based substances utilized in bone repair, such as tricalcium phosphate and HA, Bioglass® 45S5, often known as BAG, has demonstrated outstanding abilities in encouraging bone formation [94]. One noteworthy commercial product, Biogran®, primarily employed in maxillofacial procedures, features larger particle sizes ranging from 300 to 360 μm [95]. An innovative application involves mixing blood from the surgical defect with NovaBone®, another formulation based on Bioglass® 45S5, to create a putty that effectively fills the defect [96].

For addressing substantial defects, including mandibular advancements, mastoid or orbital floor fractures, and more, BonAlive® offers a particulate solution with an average particle size of 1–4 mm [97, 98]. Combining granular BAG with autogenous bone in moderate quantities has proven successful in treating significant bone lesions while reducing donor site complications significantly [99]. Additionally, StronBone®, a variant of BAG-containing SrO, is used clinically to mitigate bone resorption [100]. Clinical studies, both short-term and long-term, consistently demonstrate impressive results with BAG, showcasing excellent bone healing and reduced donor site complications [99, 100]. In dental and maxillofacial applications, FastOs®BG presents advantages over Bioglass® 45S5, offering slower resorption, superior osteoconductivity, and improved biocompatibility. This makes alkali-free BAGs excellent substitutes for Bioglass® 45S5 in these specific applications.

Craniofacial osseous reconstruction studies have explored personalized porous implants made from BAG S53P4 and fiber-reinforced composite (FCR) or poly (methyl-methacrylate) (PMMA) as supportive frameworks, yielding esthetically pleasing and functional outcomes. These studies have extended up to 4–5 years of follow-up with no adverse effects or complications [101, 102].

Personalized medicine introduces the challenge of tailoring bone substitutes to each patient's unique bone geometry. While scaffolds may not match the mechanical qualities of cranial FCR or PMMA-BAG implants, they can be utilized to deliver medications or growth hormones for specific therapeutic purposes [103, 104]. Recent advancements include the development of bioactive nanocomposite electroblown scaffolds made from polycaprolactone and BAG nanoparticles, facilitating defect filling and shaping. Additionally, human dental pulp stem cells (hDPSCs) have been successfully introduced, multiplied, and differentiated into osteogenic cells. Implantation of such scaffolds into alveolar bone defects has shown early signs of new bone growth [105].

To achieve sustained drug release in bone defects, 3D-printed scaffolds combining poly-3-hydroxybutyrate-co-3-hydroxyhexanoate (PHBHHx) and chemically modified mesoporous BAG have been used [106, 107]. In one study, isoniazid and rifampicin were incorporated to combat tuberculosis infection, resulting in enhanced differentiation of bone marrow stem cells and improved angiogenesis in a rat model [108]. The future of therapy modalities in this field may involve the use of electrospun and mesoporous BAG scaffolds, allowing for customization not only of scaffold morphology but also the inclusion of specific medications, stem cells, and growth factors to optimize individualized treatment regimens.

4.5.5 Periodontics

In the field of periodontics (figure 4.4), the development of deeper soft tissue pockets between the gingiva and tooth roots, the resorption of alveolar bone, the loss of clinical attachment level, and subsequent tooth mobility are some key indicators of

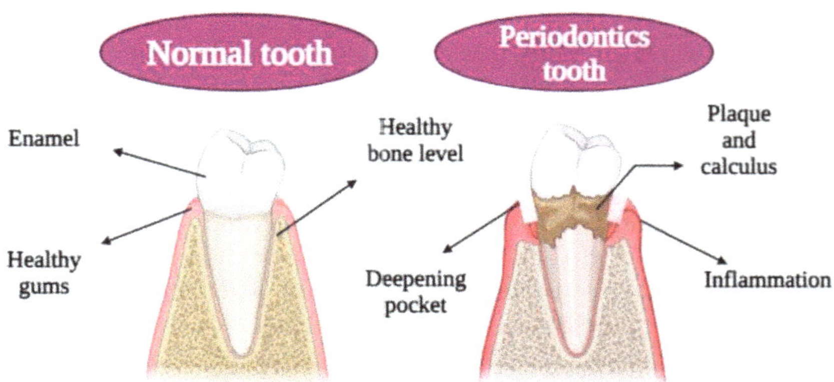

Figure 4.4. Development of periodontists along with changes to the alveolar bone density and gingiva pocket development.

periodontitis, a common chronic inflammatory disorder affecting the periodontium [109]. Additionally, periodontitis increases the risk of peri-implantitis, which might result in unstable DIs [110, 111]. Following tooth loss, it is essential to address localized alveolar ridge resorption and osseous anomalies in order to improve the prognosis for teeth and dental implants. When placing DIs, it is essential to select alveolar ridges with adequate height and bone volume. Notably, PerioGlas®, a well-known clinical brand of BAG, has played a significant role in the clinical regeneration of periodontal bone lesions. It is worth mentioning that the formulation of PerioGlas® is identical to Bioglass® 45S5 [112, 113].

To promote bone regeneration, particularly in interproximal bone defects, PerioGlas® has been widely used in periodontal surgical treatments. The benefit of doing so is that it aids in trabecular bone hemostasis [99, 112, 114]. The placement of PerioGlas® particles, which have a size range of 90–710 m, into bone defects is beneficial. Its effectiveness in treating apical osseous anomalies with endodontic surgery has also been shown by radiographic tests, leading to better success rates and quicker bone regeneration [115]. Since its inception in 1988, the Endosseous Ridge Maintenance Implant (ERMI®), a BAG product made from Bioglass® 45S5, has been used in periodontal surgery. In newly made extraction sockets, the prefabricated Bioglass® cone, ERMI®, can be inserted. With an 85.7% cone retention rate after a 5-year follow-up period, it is a secure solution for holding dentures and other dental components [116].

PerioGlas® has shown equivalent outcomes to autogenic bone grafts in randomized controlled trials (RCTs) for the treatment of grade II furcation involvement and intraosseous periodontal abnormalities, with a similar attachment gain in regeneration [117, 118]. In comparison to active control groups and open flap debridement, BAG therapy for intrabony defects significantly increases clinical attachment levels and probing depth, according to a meta-analysis [119]. Furthermore, when compared to conventional methods like closed or open debridement, BAGs have demonstrated better therapeutic success [116].

However, in order to verify true regeneration, histological examinations must be done. At the treatment location, new, functional periodontal ligaments (PDLs), alveolar bone, and cement must form in order for true regeneration to occur [120]. These traits are not present in Bioglass® 45S5 formulations since they predominantly stimulate bone production without boosting cementum or PDL, providing repair as opposed to real regeneration [121, 122]. Additionally, the granular texture of commercial BAGs used in periodontal therapy, such as PerioGlas®, makes them difficult to handle, more likely to collapse during the healing process, and less able to endure loading forces [114].

4.5.6 Enhancing dental adhesives

Dental adhesives play a pivotal role in facilitating the attachment of various substances, such as orthodontic brackets or dental composite materials, to the tooth's tissue. Overcoming the challenge of bonding hydrophobic resin composite to the hydrophilic tooth surface involves using dental resin composite to establish a strong connection between these materials. This adhesive acts as the vital link between them. However, it is important to note that the adherence of orthodontic brackets can lead to issues like white spot lesions (WSLs) and demineralization, creating favorable conditions for bacterial colonization [123]. To mitigate these concerns, preventive measures such as mouthwash, fluoride dentifrices, regular tooth brushing, or varnish applications have proven effective against WSLs. Nevertheless, these methods necessitate high levels of patient compliance and may incur additional expenses. To ensure a sustained release of fluoride during orthodontic treatment, researchers have focused on developing adhesives, primers, and fluoride-releasing sealants. It is worth noting that the addition of fluoride can compromise the mechanical properties of resin-based adhesives, leading to a gradual decrease in fluoride release over time.

When preparing a tooth for a composite restoration, a smear layer forms, mainly consisting of tooth material and bacterial byproducts. This layer coats the tooth's surface and obstructs the dentinal tubules. In order to expose the collagen network and dentinal tubules for improved infiltration by the bonding resin components, this smear layer is removed using acid-etching. However, the etchant's low pH can cause matrix metalloproteinases (MMPs) to become active and begin destroying the collagen network in dentin. As a result, poorly penetrated resin interfaces in etched dentin may cause the hybrid interface layer to deteriorate, diminish bond strength, raise the likelihood of material deterioration, and cause bond failure [124–127].

After three months of storage in simulated body fluid (SBF), the impact of two experimental resin bonding systems, one featuring Zn-polycarboxylated BAG and the other containing Bioglass® 45S5 micro-fillers, was assessed at the resin-bonded dentine interface. The increased zinc content in BAG, alongside its ability to raise pH, may have shielded the collagen network from MMP activity [128]. BAG-containing bonding systems showed a decrease in micro-permeability through the remineralization of mineral-deficient regions as compared to the BAG-free bonding system. Both 24 h and three months of SBF immersion resulted in an increase in their elasticity and hardness at the dentine contact [129].

Recently, interest has grown in a BAG-ceramic known as Biosilicate®, which is composed of Na_2O, CaO, SiO_2, and P_2O_5 in varying proportions. It has demonstrated potential in clinical bone grafting applications and when used in conjunction with titanium implants. Additionally, it has been advocated for use in total-etch adhesive bonding systems and as an alternate therapy for dentin hypersensitivity [130]. Dentinal tubule tissue reacts with Biosilicate® particles when they come into contact with dentin, causing dentinal occlusion by HA and strengthening the bond [131]. According to studies, applying a Biosilicate® suspension prior to the application of adhesive systems strengthens their binding in both mineralized and demineralized dentin [132].

The incorporation of niobophosphate BAG fillers into a commercial adhesive has yielded higher microhardness and radiopacity compared to the adhesive without BAG. Importantly, this addition did not compromise the adhesive's mechanical properties. Furthermore, apatite formation was observed [133]. An innovative BAG-resin orthodontic adhesive, containing fluoride, has shown potential in promoting apatite formation under neutral and acidic conditions. This suggests a clinical role in remineralization and the prevention of WSLs around orthodontic brackets.

Commercial orthodontic bonding solutions with flowable resin with BAG, Ag, or Zn-doped fillers have shown promising results. After pH cycling, the inclusion of BAG produced a demineralization-free zone that reached up to 200–300 m from the bracket. Surfaces that weren't coated in BAG-free bonding agents (used as controls) were demineralized in comparison. Additionally, compared to the controls, these experimental bonding agents significantly inhibited *S. mutans*, suggesting their potential use in orthodontic procedures [134].

4.5.7 Restorative materials

Presently available restorative materials can closely mimic the appearance, structure, and function of natural teeth but lack bioactive attributes. Both resin composite materials and glass ionomer cement experience some level of polymerization shrinkage when used for cavity restoration [135]. This can result in the formation of microgaps due to differences in the mechanical properties between natural tooth and the restorative material, potentially leading to a common cause of dental restoration failure, secondary caries, which is often challenging to address through conventional dental care methods [136]. Additionally, the tissue-saving techniques used to remove caries may inadvertently leave behind residual bacteria in the affected tissue [137]. Figure 4.5 represents the BAG as a restorative material in the dentistry field.

The creation of restorative materials capable of remineralizing or fixing demineralized dentin after bacterial invasion has been a major focus of dental biomaterial research. Dental restorations can be made to survive for a very long time by forming a solid bond with the tooth and producing an unfavorable environment for microorganisms. The creation of a sealed interface by the precipitation of HA can

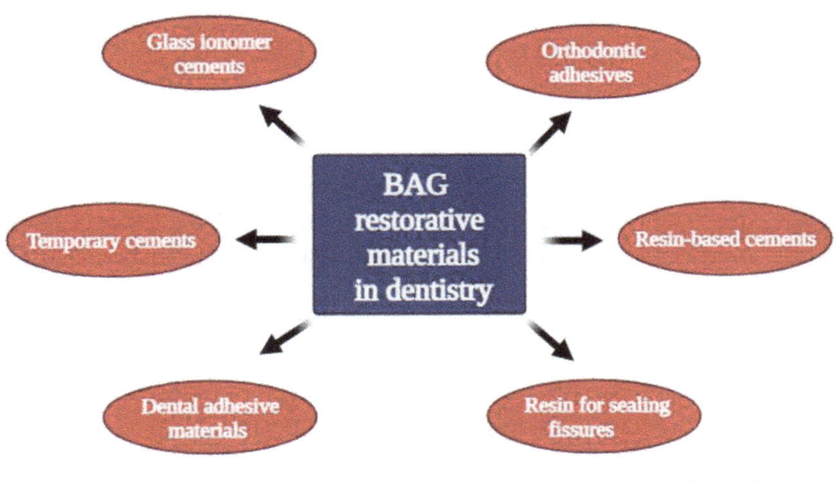

Figure 4.5. Schematic representation of BAGs as restorative materials in dentistry.

be facilitated by the use of bonding agents having bioactive characteristics [138]. Notably, it has been demonstrated that Bioglass® 45S5 promotes dentin remineralization. After being exposed to bacterial challenges and aqueous circumstances for two months, it showed better mechanical capabilities compared to the control (0 weight percent Bioglass® 45S5) when added to a resin composite at weight percentages of 5, 10, and 15 with a filler content of 72% [139]. In contrast to commercially available resin composites, these Bioglass® composites did, however, exhibit cytotoxicity because unreacted monomers were released into the environment [140]. However, flowable resin composite materials have shown to have the strength of the bond without losing the capacity to stop the growth of oral bacteria such *S. mutans* and *Escherichia coli* [141].

By using resin composites containing Bioglass® (BAG) and fluoride, dentin remineralization can be hastened and enzymatic deterioration at the dentin interface can be reduced. In samples exposed to artificial saliva for 3 and 30 days, respectively, the ability of F-BAG (fluoride-modified BAG) and BAG resin composites to promote remineralization was assessed. F-BAG demonstrated the best rate of dentin remineralization and the least amount of enzymatic destruction of the dentin collagen network, suggesting possible benefits over traditional BAG, 45S5 [22].

Ag-doped BAG resin composites were also tested for their bioactivity and antibacterial qualities. In comparison to control samples (BAG-free resin composite), the concentration of Ag-BAG resin composite increased the number of dead bacteria in the biofilm and improved the development of apatite while maintaining identical mechanical qualities. These results imply that Ag-BAG resin composites may be highly effective in reducing the development of secondary caries [142].

BAG resin composites can have different mechanical characteristics. When immersed in brain heart infusion media for two months, experimental resin composites with 0–15 weight percent of the fillers replaced by ground BAG remained stable and displayed similar mechanical characteristics, with the exception of reduced fracture toughness and fatigue resistance, compared to three commercial composites [139]. As the weight percent of BAG grew, the dentin bond strength in a resin composite with different levels of BAG (0–40 weight percent) and 70–% filler content linearly decreased [143]. After artificial ageing in ethanol and water, identical experimental BAG resin composites were assessed for flexural strength, flexural modulus, modulus of resilience, and material dependability. Flexural strength and modulus were reduced further by artificial ageing and decreased linearly with increasing BAG content. Up to 20 weight percent of BAG, ISO 4049 criteria for minimum flexural strength were reached; however, as BAG was added, the modulus of resilience and degree of conversion dropped [144]. A dose-dependent decrease in flexural strength and an increase in water sorption were seen when BAG content rose in experimental pit and fissure sealants with varied BAG contents (0–50 weight percent) [145]. However, experimental composites with up to 15% weight-per-percent sodium-free BAG and up to 72% weight-per-percent reinforcing fillers showed comparable flexural strength to BAG-free resin composites [139].

While adding BAG filler particles to resin composites can give them antibacterial and bioactive qualities that are good for avoiding secondary caries, doing so may damage their mechanical, optical, or adhesive capabilities [139]. Commonly utilized in conventional resin composites to improve mechanical qualities, silanization of filler particles may decrease ion release and consequently affect bioactivity [146, 147]. Reducing resin hydrophilicity may lessen the composite's capacity to remineralize minerals while increasing ageing resistance. These inconsistent results highlight how difficult it is to maintain a balance between mechanical characteristics and deterioration in BAG resin composites.

Glass ionomer cement (GIC) is generally made of fluoride-aluminosilicate glass and polyacrylic acid, but it can be altered to improve its mechanical qualities, adhesion, and solubility by adding methacrylate resin monomers (rmGIC). GIC is renowned for its ability to release fluoride, promote remineralization, and form a strong chemical link with teeth. To control remineralization, researchers have added BAG particles to GIC formulations [148, 149]. An acid–base reaction between glass particles and polyacrylic acid during setting was seen in a GIC made of BAG and polyacrylic acid [150]. According to the theory, as an iron-rich matrix develops during GIC setup, an osmotic gradient is created that enables water to be absorbed by the matrix, providing an aqueous environment where BAG particles can react.

Additionally, it has been demonstrated that rmGICs absorb more water, which supports the higher bioactivity of BAG-rmGIC in comparison to conventional GIC [151]. Higher flexural strength in demineralized dentin submerged in BAG-rmGIC-containing solutions showed that BAG-rmGIC also exhibited superior remineralizing properties versus rmGIC [151]. BAG-rmGIC with 10–30 weight percent BAG showed excellent results for remineralization in an *in vivo* investigation including class III restorations in undamaged beagle dog teeth, with calcium phosphate deposition seen on the restoration surface and in deeper areas of the dentinal tubules [152]. BAG alone provided antimicrobial activities against both *S. mutans* and *Candida albicans*, however, *in vitro* tests with BAG-GICs with 10–30 weight percent BAG (S53P) demonstrated antimicrobial properties against both pathogens at 30 weight percent BAG [153]. These results support BAG-GICs' potential clinical use in reducing secondary caries. It is important to note that whereas high concentrations of nanoparticle BAG (nBAG) in rmGICs had a negative impact on mechanical parameters due to reduced bonding between glass particles and the resin matrix, low concentrations had a positive impact on flexural strength [154]. The integration of BAG particles into GICs may degrade their mechanical characteristics despite their bioactive and antibacterial qualities, limiting their clinical usage to regions with minimal mechanical loads that require bioactivity, such as root surface fillings and liners [155].

4.5.8 Dentin hypersensitivity

DH is defined by its hallmark symptoms of acute, brief dental pain triggered by various stimuli, such as touch, chemicals, osmotic changes, evaporation, or temperature fluctuations. DH can arise when the dentin is exposed due to erosion, attrition, abfraction, abrasion, gingival recession, or periodontal disease. The prevailing theory explaining DH is the hydrodynamic theory, which posits that these stimuli initiate fluid movement within the dentinal tubules. This, in turn, stimulates mechanoreceptors near the pulp, exciting the nerve terminals of $A\delta$ fibers and resulting in the characteristic pain perception [156, 157]. According to the hydrodynamic theory, DH can be managed either by inhibiting nerve excitation or by sealing the dentinal tubules. One approach to blocking excitation is by increasing the extracellular concentration of potassium ions around nerve fibers, thereby preventing action potential generation [158]. Occluding open dentinal tubules reduces dentinal fluid flow [159].

Over-the-counter products used in the conservative management of DH include GIC, bonding agents, and dentifrices [18]. Some of these products incorporate glass particles, and formulations involving BAG can provide therapeutic relief by sealing the dentinal tubules through binding to collagen fibers and depositing HA [160]. Novamin®, introduced in 2004 with a particle size of 18 μm as an ingredient in Sensodyne® toothpaste, is one such product designed to treat DH [161]. Sensodyne® has gained widespread endorsement from dentists [18]. PerioGlas® has also shown success in treating DH, primarily due to its strong affinity for collagen, which facilitates dentin bonding and tubule occlusion [162]. The quantity of BAG used correlates with the degree of tubule occlusion, with BAG replacing silica in toothpaste offering greater resistance to pH rinses and brushing [18]. BAG particles synthesized via the sol–gel technique exhibit increased surface area and rapid bonding to dentin compared to melt-quenched glass.

Despite the abundance of *in vitro* evidence, limited *in vivo* data exist regarding the clinical effectiveness of Novamin® in treating DH. Gendreau *et al* conducted a review of available clinical studies, lending support to the clinical effectiveness of Novamin® when used in toothpaste [163]. In a comparison with a dentifrice formulation containing 5% potassium nitrate (5% KNO3), a 5% Novamin® formulation demonstrated significantly lower pain scores and longer-lasting relief [164]. Toothpaste containing fluoride-containing BAGs, which can form FAp, is suggested to offer a more effective treatment for DH [165]. BioMin-F®, which forms FAp, has been subjected to an RCT to evaluate its clinical desensitizing properties in comparison to NovaMin® and a standard fluoride toothpaste.

4.6 Conclusion

BAGs have emerged as remarkable materials with wide-ranging applications in dentistry. These non-crystalline ceramics possess unique properties that make them highly attractive for tissue engineering and implant construction. BAGs like 45S5 Bioglass® exhibit a convergence of mechanical biocompatibility and physiological activity, offering new possibilities in healthcare. They bond with living tissues, stimulate new tissue formation, and gradually degrade over time. Initially designed for bone defect treatment, BAGs have expanded to include dental coatings, bone grafts, restorative materials, and tissue engineering scaffolds. In dentistry, BAGs find versatile applications from dental implant coatings to bone grafts for osteogenesis. They promote bone growth and interact with biological systems, making them valuable for improving patient quality of life. BAGs represent an intriguing intersection of materials science, medicine, and dentistry. Their versatility, bioactivity, and antimicrobial properties hold great promise for tissue engineering, implant construction, and oral healthcare. Ongoing research is crucial to unlock BAGs' full potential and enhance their clinical applications.

References

[1] Jones J 2015 Reprint of: review of bioactive glass: from Hench to hybrids *Acta Biomater.* **23** S53–82

[2] Kaur G, Pandey O P, Singh K, Homa D, Scott B and Pickrell G 2014 A review of bioactive glasses: their structure, properties, fabrication and apatite formation *J. Biomed. Mater. Res. Part* A **102** 254–74

[3] Javed F, Vohra F, Zafar S and Almas K 2014 Significance of osteogenic surface coatings on implants to enhance osseointegration under osteoporotic-like conditions *Implant Dent.* **23** 679–86

[4] Najeeb S, Bds Z K, Bds S Z and Bds M S 2016 Bioactivity and osseointegration of PEEK are inferior to those of titanium: a systematic review *J. Oral Implantol.* **42** 512–6

[5] Wahaj A, Hafeez K and Zafar M S 2016 Role of bone graft materials for cleft lip and palate patients: a systematic review *Saudi J. Dental Res.* **7** 57–63

[6] Morselli P G, Giuliani R, Pinto V, Oranges C M, Negosanti L, Tavaniello B and Morellini A 2009 Treatment of alveolar cleft performing a pyramidal pocket and an autologous bone grafting *J. Craniofac. Surg.* **20** 1566–70

[7] Najeeb S, Khurshid Z, Zafar M S, Khan A S, Zohaib S, Martí J M, Sauro S, Matinlinna J P and Rehman I U 2016 Modifications in glass ionomer cements: nano-sized fillers and bioactive nanoceramics *Int. J. Mol. Sci.* **17** 1134

[8] Zafar M S and Ahmed N 2015 Therapeutic roles of fluoride released from restorative dental materials *Fluoride.* **48** 184–94

[9] Zafar M, Najeeb S, Khurshid Z, Vazirzadeh M, Zohaib S, Najeeb B and Sefat F 2016 Potential of electrospun nanofibers for biomedical and dental applications *Materials* **9** 73

[10] Zafar M S, Khurshid Z and Almas K 2015 Oral tissue engineering progress and challenges *Tissue Eng. Regen. Med.* **12** 387–97

[11] Bekmurzayeva A, Duncanson W J, Azevedo H S and Kanayeva D 2018 Surface modification of stainless steel for biomedical applications: revisiting a century-old material *Mater. Sci. Eng.* C **93** 1073–89

[12] Lemons J E and Lucas L C 1986 Properties of biomaterials *J. Arthroplasty* **1** 143–7

[13] Wahid F, Khan T, Hussain Z and Ullah H 2018 Nanocomposite scaffolds for tissue engineering; properties, preparation and applications *Applications of Nanocomposite Materials in Drug Delivery* (Cambridge: Woodhead Publishing) pp 701–35

[14] Hajiali H, Karbasi S, Hosseinalipour M and Rezaie H R 2010 Preparation of a novel biodegradable nanocomposite scaffold based on poly (3-hydroxybutyrate)/bioglass nanoparticles for bone tissue engineering *J. Mater. Sci., Mater. Med.* **21** 2125–32

[15] Jones J 2013 Review of bioactive glass: from Hench to hybrids *Acta Biomater.* **9** 4457–86

[16] Albrektsson T and Johansson C 2001 Osteoinduction, osteoconduction and osseointegration *Eur. Spine. J.* **10** S96–101

[17] Hench L L 2006 The story of Bioglass® *J. Mater. Sci., Mater. Med.* **17** 967–78

[18] Gillam D G, Bulman J S, Eijkman M A and Newman H N 2002 Dentists' perceptions of dentine hypersensitivity and knowledge of its treatment *J. Oral Rehabil.* **29** 219–25

[19] Earl J S, Ward M B and Langford R M 2010 Investigation of dentinal tubule occlusion using FIB-SEM milling and EDX *J. Clin. Dent.* **21** 37–41
[20] Greenspan D C 2010 NovaMin® and tooth sensitivity—an overview *J. Clin. Dent.* **21** 61
[21] Earl J S, Leary R K, Muller K H, Langford R M and Greenspan D C 2011 Physical and chemical characterization of dentin surface following treatment with NovaMin technology *J. Clin. Dent.* **22** 62–7
[22] Tezvergil-Mutluay A, Seseogullari-Dirihan R, Feitosa V P, Cama G, Brauer D S and Sauro S 2017 Effects of composites containing bioactive glasses on demineralized dentin *J. Dent. Res.* **96** 999–1005
[23] Hench L L, Spilman D B and Hench J W 1988 Inventors; University of Florida, assignee. Fluoride-containing Bioglass™ compositions *United States Patent US* 4,775,646
[24] Andersson Ö H, Liu G, Karlsson K H, Niemi L, Miettinen J and Juhanoja J 1990 *In vivo* behaviour of glasses in the SiO_2–Na_2O–CaO–P_2O_5–Al_2O_3–B_2O_3 system *J. Mater. Sci., Mater. Med.* **1** 219–27
[25] Al-Eesa N A, Johal A, Hill R G and Wong F S 2018 Fluoride containing bioactive glass composite for orthodontic adhesives—apatite formation properties *Dent. Mater.* **34** 1127–33
[26] Bakry A S, Abbassy M A, Alharkan H F, Basuhail S, Al-Ghamdi K and Hill R 2018 A novel fluoride containing bioactive glass paste is capable of re-mineralizing early caries lesions *Materials* **11** 1636
[27] Rheinberger V, Salz U, Höland W, Rumphorst A, Grabher K, Fischer U K, Schweiger M and Moszner N 2002 Inventors; Ivoclar AG, assignee. Polymerizable composite material *United States Patent US* 6,353,039
[28] Brauer D S, Karpukhina N, Law R V and Hill R G 2009 Structure of fluoride-containing bioactive glasses *J. Mater. Chem.* **19** 5629–36
[29] Pajor K, Pajchel L and Kolmas J 2019 Hydroxyapatite and fluorapatite in conservative dentistry and oral implantology—a review *Materials* **12** 2683
[30] Mneimne M, Hill R G, Bushby A J and Brauer D S 2011 High phosphate content significantly increases apatite formation of fluoride-containing bioactive glasses *Acta Biomater.* **7** 1827–34
[31] Goldstep F 2012 Proactive intervention dentistry: a model for oral care through life *Compend. Contin. Educ. Dent.* **33** 394–6
[32] Chen Q Z, Xu J L, Yu L G, Fang X Y and Khor K A 2012 Spark plasma sintering of sol–gel derived 45S5 Bioglass®-ceramics: mechanical properties and biocompatibility evaluation *Mater. Sci. Eng.* C **32** 494–502
[33] Filho O P, La Torre G P and Hench L L 1996 Effect of crystallization on apatite-layer formation of bioactive glass 45S5 *J. Biomed. Mater. Res.* **30** 509–14
[34] Prasad S, Vyas V K, Ershad M and Pyare R 2017 Crystallization and mechanical properties of (45S5-HA) biocomposite for biomedical implantation *J. Ceram.-Silk.* **61** 378–84
[35] Wilson J, Pigott G H, Schoen F J and Hench L L 1981 Toxicology and biocompatibility of bioglasses *J. Biomed. Mater. Res.* **15** 805–17
[36] Fernandes H R, Gaddam A, Rebelo A, Brazete D, Stan G E and Ferreira J M 2018 Bioactive glasses and glass-ceramics for healthcare applications in bone regeneration and tissue engineering *Materials* **11** 2530
[37] Midha S, Kim T B, Van Den Bergh W, Lee P D, Jones J R and Mitchell C A 2013 Preconditioned 70S30C bioactive glass foams promote osteogenesis *in vivo Acta Biomater.* **9** 9169–82

[38] Wu C, Fan W, Gelinsky M, Xiao Y, Simon P, Schulze R, Doert T, Luo Y and Cuniberti G 2011 Bioactive SrO–SiO2 glass with well-ordered mesopores: characterization, physiochemistry and biological properties *Acta Biomater.* **7** 1797–806
[39] Palmer L C, Newcomb C J, Kaltz S R, Spoerke E D and Stupp S I 2008 Biomimetic systems for hydroxyapatite mineralization inspired by bone and enamel *Chem. Rev.* **108** 4754–83
[40] Vichery C and Nedelec J M 2016 Bioactive glass nanoparticles: from synthesis to materials design for biomedical applications *Materials* **9** 288
[41] Sepulveda P, Jones J R and Hench L L 2002 *In vitro* dissolution of melt-derived 45S5 and sol–gel derived 58S bioactive glasses *J. Biomed. Mater. Res.* **61** 301–11
[42] Vogel M, Voigt C, Gross U M and Müller-Mai C M 2001 *In vivo* comparison of bioactive glass particles in rabbits *Biomaterials* **22** 357–62
[43] Damen J J and Ten Cate J M 1992 Silica-induced precipitation of calcium phosphate in the presence of inhibitors of hydroxyapatite formation *J. Dent. Res.* **71** 453–7
[44] Salonen J I, Arjasmaa M, Tuominen U, Behbehani M J and Zaatar E I 2009 Bioactive glass in dentistry *J. Minim. Interv. Dent.* **2** 208–18
[45] Ali S, Farooq I and Iqbal K 2014 A review of the effect of various ions on the properties and the clinical applications of novel bioactive glasses in medicine and dentistry *Saudi Dent. J.* **26** 1–5
[46] Chen Q Z, Thompson I D and Boccaccini A R 2006 45S5 Bioglass®-derived glass–ceramic scaffolds for bone tissue engineering *Biomaterials* **27** 2414–25
[47] Chen Q, Baino F, Spriano S, Pugno N M and Vitale-Brovarone C 2014 Modelling of the strength–porosity relationship in glass-ceramic foam scaffolds for bone repair *J. Eur. Ceram. Soc.* **34** 2663–73
[48] Govender S, Lutchman D, Pillay V, Chetty D J and Govender T 2006 Enhancing drug incorporation into tetracycline-loaded chitosan microspheres for periodontal therapy *J. Microencapsulation* **23** 750–61
[49] Allan I, Newman H and Wilson M 2001 Antibacterial activity of particulate Bioglass® against supra-and subgingival bacteria *Biomaterials* **22** 1683–7
[50] Zhang D, Leppäranta O, Munukka E, Ylänen H, Viljanen M K, Eerola E, Hupa M and Hupa L 2010 Antibacterial effects and dissolution behavior of six bioactive glasses *J. Biomed. Mater. Res.* A **93** 475–83
[51] Zhang X I, Jia W T, Gu Y F, Zhang C Q, Huang W H and Wang D P 2010 Borate bioglass based drug delivery of teicoplanin for treating osteomyelitis *J. Inorg. Mater.* **25** 293–8
[52] Gomez-Vega J M, Saiz E, Tomsia A P, Marshall G W and Marshall S J 1999 Bioactive glass coatings with hydroxyapatite and Bioglass (registered) particles on Ti-based implants. 1. Processing *Biomaterials* **21** 105–11
[53] Najeeb S, Zafar M S, Khurshid Z and Siddiqui F 2016 Applications of polyetheretherketone (PEEK) in oral implantology and prosthodontics *J. Prosthodont. Res.* **60** 12–9
[54] Polo-Corrales L, Latorre-Esteves M and Ramirez-Vick J E 2014 Scaffold design for bone regeneration *J. Nanosci. Nanotechnol.* **14** 15–56
[55] Välimäki V V and Aro H T 2006 Molecular basis for action of bioactive glasses as bone graft substitute *Scand. J. Surg.* **95** 95–102
[56] Oonishi H, Kushitani S, Yasukawa E, Iwaki H, Hench L L, Wilson J, Tsuji E and Sugihara T 1997 Particulate bioglass compared with hydroxyapatite as a bone graft substitute *Clin. Orthop. Relat. Res.* **334** 316–25

[57] van Gestel N A, Geurts J, Hulsen D J, van Rietbergen B, Hofmann S and Arts J J 2015 Clinical applications of S53P4 bioactive glass in bone healing and osteomyelitic treatment: a literature review *BioMed Res. Int.* **2015** 684826

[58] Gerhardt L C and Boccaccini A R 2010 Bioactive glass and glass-ceramic scaffolds for bone tissue engineering *Materials* **3** 3867–910

[59] Brydone A S, Meek D and Maclaine S 2010 Bone grafting, orthopaedic biomaterials, and the clinical need for bone engineering *Proc. Inst. Mech. Eng. Part H: J. Eng. Med.* **224** 1329–43

[60] Janicki P and Schmidmaier G 2011 What should be the characteristics of the ideal bone graft substitute? Combining scaffolds with growth factors and/or stem cells *Injury* **42** S77–81

[61] Arrington E D, Smith W J, Chambers H G, Bucknell A L and Davino N A 1996 Complications of iliac crest bone graft harvesting *Clin. Orthop. Relat. Res.* **329** 300–9

[62] Banwart J C, Asher M A and Hassanein R S 1995 Iliac crest bone graft harvest donor site morbidity: a statistical evaluation *Spine* **20** 1055–60

[63] Kinaci A, Neuhaus V and Ring D C 2014 Trends in bone graft use in the United States *Orthopedics* **37** e783–8

[64] Müller F, Wahl G and Fuhr K 1994 Age-related satisfaction with complete dentures, desire for improvement and attitudes to implant treatment *Gerodontology* **11** 7–12

[65] Albrektsson T, Brånemark P I, Hansson H A and Lindström J 1981 Osseointegrated titanium implants: requirements for ensuring a long-lasting, direct bone-to-implant anchorage in man *Acta Orthop. Scand.* **52** 155–70

[66] Yeo I S, Kim H Y, Lim K S and Han J S 2012 Implant surface factors and bacterial adhesion: a review of the literature *Int. J. Artif. Organs* **35** 762–72

[67] Talreja P S, Gayathri G V and Mehta D S 2013 Treatment of an early failing implant by guided bone regeneration using resorbable collagen membrane and bioactive glass *J. Indian Soc. Periodontol.* **17** 131

[68] Alghamdi H S 2018 Methods to improve osseointegration of dental implants in low quality (type-IV) bone: an overview *J. Funct. Biomater.* **9** 7

[69] Petersen R C 2014 Titanium implant osseointegration problems with alternate solutions using epoxy/carbon-fiber-reinforced composite *Metals* **4** 549–69

[70] Wennerberg A, Bougas K, Jimbo R and Albrektsson T 2013 Implant coatings: new modalities for increased osseointegration *Am. J. Dent.* **26** 105–12

[71] MacDonald D E, Betts F, Doty S B and Boskey A L 2000 A methodological study for the analysis of apatite-coated dental implants retrieved from humans *Ann. Periodontol.* **5** 175–84

[72] Civantos A, Martínez-Campos E, Ramos V, Elvira C, Gallardo A and Abarrategi A 2017 Titanium coatings and surface modifications: toward clinically useful bioactive implants *ACS Biomater. Sci. Eng.* **3** 1245–61

[73] Mistry S, Kundu D, Datta S and Basu D 2011 Comparison of bioactive glass coated and hydroxyapatite coated titanium dental implants in the human jaw bone *Aust. Dental J.* **56** 68–75

[74] Verné E 2012 Bioactive glass and glass-ceramic coatings *Bio-Glasses—An Introduction* (Wiley) pp 107–19

[75] Lopez-Esteban S, Saiz E, Fujino S, Oku T, Suganuma K and Tomsia A P 2003 Bioactive glass coatings for orthopedic metallic implants *J. Eur. Ceram. Soc.* **23** 2921–30

[76] Fujino S, Tokunaga H, Saiz E and Tomsia A P 2004 Fabrication and characterization of bioactive glass coatings on Co-Cr implant alloys *Mater. Trans.* **45** 1147–51

[77] Monsalve M, Ageorges H, Lopez E, Vargas F and Bolivar F 2013 Bioactivity and mechanical properties of plasma-sprayed coatings of bioglass powders *Surf. Coat. Technol.* **220** 60–6

[78] López Calvo V, Vicent Cabedo M, Bannier E, Cañas Recacha E, Boccaccini A R, Cordero Arias L and Sánchez Vilches E 2014 45S5 bioactive glass coatings by atmospheric plasma spraying obtained from feedstocks prepared by different routes *J. Mater. Sci.* **49** 7933–42

[79] Fu T, Alajmi Z, Shen Y, Wang L, Yang S and Zhang M 2017 Sol–gel preparation and properties of Ag-containing bioactive glass films on titanium *Int. J. Appl. Ceram. Technol.* **14** 1117–24

[80] Hamadouche M, Meunier A, Greenspan D C, Blanchat C, Zhong J P, La Torre G P and Sedel L 2000 Bioactivity of sol-gel bioactive glass coated alumina implants *J. Biomed. Mater. Res.* **52** 422–9

[81] Xue B, Guo L, Chen X, Fan Y, Ren X, Li B, Ling Y and Qiang Y 2017 Electrophoretic deposition and laser cladding of bioglass coating on Ti *J. Alloys Compd.* **710** 663–9

[82] Krause D, Thomas B, Leinenbach C, Eifler D, Minay E J and Boccaccini A R 2006 The electrophoretic deposition of Bioglass® particles on stainless steel and Nitinol substrates *Surf. Coat. Technol.* **200** 4835–45

[83] Popescu A C, Sima F, Duta L, Popescu C, Mihailescu I N, Capitanu D, Mustata R, Sima L E, Petrescu S M and Janackovic D 2009 Biocompatible and bioactive nanostructured glass coatings synthesized by pulsed laser deposition: *in vitro* biological tests *Appl. Surf. Sci.* **255** 5486–90

[84] D'Alessio L, Teghil R, Zaccagnino M, Zaccardo I, Ferro D and Marotta V 1999 Pulsed laser ablation and deposition of bioactive glass as coating material for biomedical applications *Appl. Surf. Sci.* **138** 527–32

[85] Wang C X, Chen Z Q and Wang M 2002 Fabrication and characterization of bioactive glass coatings produced by the ion beam sputter deposition technique *J. Mater. Sci., Mater. Med.* **13** 247–51

[86] Popa A C, Stan G E, Husanu M A, Mercioniu I, Santos L F, Fernandes H R and Ferreira J M 2017 Bioglass implant-coating interactions in synthetic physiological fluids with varying degrees of biomimicry *Int. J. Nanomed.* **2017** 683–707

[87] Mardare C C, Mardare A I, Fernandes J R, Joanni E, Pina S C, Fernandes M H and Correia R N 2003 Deposition of bioactive glass-ceramic thin-films by RF magnetron sputtering *J. Eur. Ceram. Soc.* **23** 1027–30

[88] Stan G E, Morosanu C O, Marcov D A, Pasuk I, Miculescu F and Reumont G 2009 Effect of annealing upon the structure and adhesion properties of sputtered bio-glass/titanium coatings *Appl. Surf. Sci.* **255** 9132–8

[89] Šimek M, Černák M, Kylián O, Foest R, Hegemann D and Martini R 2019 White paper on the future of plasma science for optics and glass *Plasma Process. Polym.* **16** 1700250

[90] Wasa K, Kitabatake M and Adachi H 2004 *Thin Film Materials Technology: Sputtering of Control Compound Materials* (Berlin: Springer Science & Business Media)

[91] Moritz N, Rossi S, Vedel E, Tirri T, Ylänen H, Aro H and Närhi T 2004 Implants coated with bioactive glass by CO 2-laser, an *in vivo* study *J. Mater. Sci., Mater. Med.* **15** 795–802

[92] Wheeler D L, Montfort M J and McLoughlin S W 2001 Differential healing response of bone adjacent to porous implants coated with hydroxyapatite and 45S5 bioactive glass *J. Biomed. Mater. Res.* **55** 603–12
[93] Popa A C, Stan G E, Enculescu M, Tanase C, Tulyaganov D U and Ferreira J M 2015 Superior biofunctionality of dental implant fixtures uniformly coated with durable bioglass films by magnetron sputtering *J. Mech. Behav. Biomed. Mater.* **51** 313–27
[94] Peltola M J, Aitasalo K M, Suonpää J T, Yli-Urpo A, Laippala P J and Forsback A P 2003 Frontal sinus and skull bone defect obliteration with three synthetic bioactive materials. A comparative study *J. Biomed. Mater. Res.* B **66** 364–72
[95] Tadjoedin E S, De Lange G L, Lyaruu D M, Kuiper L and Burger E H 2002 High concentrations of bioactive glass material (BioGran®) vs. autogenous bone for sinus floor elevation: histomorphometrical observations on three split mouth clinical cases *Clin. Oral Implants Res.* **13** 428–36
[96] Hench L L, Hench J W and Greenspan D C 2004 Bioglass: a short history and bibliography *J. Australas. Ceram. Soc.* **40** 1–42
[97] Gosain A K 2004 Plastic Surgery Educational Foundation DATA Committee. Bioactive glass for bone replacement in craniomaxillofacial reconstruction *Plast. Reconstruct. Surg.* **114** 590–3
[98] Peltola M, Aitasalo K, Suonpää J, Varpula M and Yli-Urpo A 2006 Bioactive glass S53P4 in frontal sinus obliteration: a long-term clinical experience *Head Neck* **28** 834–41
[99] Profeta A and Huppa C 2016 Bioactive-glass in oral and maxillofacial surgery *Craniomax. Trauma Rec* **9** 1–14
[100] Fujikura K, Karpukhina N, Kasuga T, Brauer D S, Hill R G and Law R V 2012 Influence of strontium substitution on structure and crystallisation of Bioglass® 45S5 *J. Mater. Chem.* **22** 7395–402
[101] Aitasalo K M, Piitulainen J M, Rekola J and Vallittu P K 2014 Craniofacial bone reconstruction with bioactive fiber-reinforced composite implant *Head Neck* **36** 722–8
[102] Peltola M J, Vallittu P K, Vuorinen V, Aho A A, Puntala A and Aitasalo K M 2012 Novel composite implant in craniofacial bone reconstruction *Eur. Arch. Oto-Rhino-Laryngol.* **269** 623–8
[103] Hum J and Boccaccini A R 2012 Bioactive glasses as carriers for bioactive molecules and therapeutic drugs: a review *J. Mater. Sci., Mater. Med.* **23** 2317–33
[104] Vallet-Regí M, Balas F and Arcos D 2007 Mesoporous materials for drug delivery *Angew. Chem. Int. Ed.* **46** 7548–58
[105] Mandakhbayar N, El-Fiqi A, Dashnyam K and Kim H W 2018 Feasibility of defect tunable bone engineering using electroblown bioactive fibrous scaffolds with dental stem cells *ACS Biomater. Sci. Eng.* **4** 1019–28
[106] Zhao S, Zhu M, Zhang J, Zhang Y, Liu Z, Zhu Y and Zhang C 2014 Three dimensionally printed mesoporous bioactive glass and poly (3-hydroxybutyrate-co-3-hydroxyhexanoate) composite scaffolds for bone regeneration *J. Mater. Chem.* B **2** 6106–18
[107] Zhu M, Li K, Zhu Y, Zhang J and Ye X 2015 3D-printed hierarchical scaffold for localized isoniazid/rifampin drug delivery and osteoarticular tuberculosis therapy *Acta Biomater.* **16** 145–55
[108] Wu C, Zhou Y, Chang J and Xiao Y 2013 Delivery of dimethyloxallyl glycine in mesoporous bioactive glass scaffolds to improve angiogenesis and osteogenesis of human bone marrow stromal cells *Acta Biomater.* **9** 9159–68

[109] Hirschfeld L and Wasserman B 1978 A long-term survey of tooth loss in 600 treated periodontal patients *J. Periodontol.* **49** 225–37

[110] Heitz-Mayfield L J 2008 Peri-implant diseases: diagnosis and risk indicators *J. Clin. Periodontol.* **35** 292–304

[111] Renvert S and Persson G R 2009 Periodontitis as a potential risk factor for peri-implantitis *J. Clin. Periodontol.* **36** 9–14

[112] Lovelace T B, Mellonig J T, Meffert R M, Jones A A, Nummikoski P V and Cochran D L 1998 Clinical evaluation of bioactive glass in the treatment of periodontal osseous defects in humans *J. Periodontol.* **69** 1027–35

[113] Singh M P and Mehta D S 2000 Clinical evaluation of Biogran as a graft material in the treatment of periodontal osseous defects *J. Indian Soc. Periodontol.* **3** 69–72

[114] Profeta A C and Prucher G M 2015 Bioactive-glass in periodontal surgery and implant dentistry *Dental Mater. J.* **34** 559–71

[115] Pantchev A, Nohlert E and Tegelberg Å 2009 Endodontic surgery with and without inserts of bioactive glass PerioGlas®—a clinical and radiographic follow-up *Oral Maxillofac. Surg.* **13** 21–6

[116] Stanley H R, Hall M B, Clark A E, King C J, Hench L L and Berte J J 1997 Using 45S5 bioglass cones as endosseous ridge maintenance implants to prevent alveolar ridge resorption: a 5-year evaluation *Int. J. Oral Maxillofac. Implants* **12** 95–105

[117] Sumer M, Keles G C, Cetinkaya B O, Balli U, Pamuk F and Uckan S 2013 Autogenous cortical bone and bioactive glass grafting for treatment of intraosseous periodontal defects *Eur. J. Dent.* **7** 6–14

[118] Skallevold H E, Rokaya D, Khurshid Z and Zafar M S 2019 Bioactive glass applications in dentistry *Int. J. Mol. Sci.* **20** 5960

[119] Sohrabi K, Saraiya V, Laage T A, Harris M, Blieden M and Karimbux N 2012 An evaluation of bioactive glass in the treatment of periodontal defects: a meta-analysis of randomized controlled clinical trials *J. Periodontol.* **83** 453–64

[120] Shue L, Yufeng Z and Mony U 2012 Biomaterials for periodontal regeneration: a review of ceramics and polymers *Biomatter* **2** 271–7

[121] Nevins M L, Camelo M, Nevins M, King C J, Oringer R J, Schenk R K and Fiorellini J P 2000 Human histologic evaluation of bioactive ceramic in the treatment of periodontal osseous defects *Int. J. Periodontics Restorative Dent.* **20** 458–67

[122] Sculean A, Windisch P, Keglevich T and Gera I 2005 Clinical and histologic evaluation of an enamel matrix protein derivative combined with a bioactive glass for the treatment of intrabony periodontal defects in humans *Int. J. Periodontics Restorative Dent.* **25** 139–47

[123] Gange P 2015 The evolution of bonding in orthodontics *Am. J. Orthod. Dentofacial Orthop.* **147** S56–63

[124] Tay F R, Pashley D H, Yiu C, Cheong C, Hashimoto M, Itou K, Yoshiyama M and King N M 2004 Nanoleakage types and potential implications: evidence from unfilled and filled adhesives with the same resin composition *Am. J. Dent.* **17** 182–90

[125] Hashimoto M, Tay F R, Ohno H, Sano H, Kaga M, Yiu C, Kumagai H, Kudou Y, Kubota M and Oguchi H 2003 SEM and TEM analysis of water degradation of human dentinal collagen *J. Biomed. Mater. Res.* B **66** 287–98

[126] Mazzoni A, Pashley D H, Nishitani Y, Breschi L, Mannello F, Tjäderhane L, Toledano M, Pashley E L and Tay F R 2006 Reactivation of inactivated endogenous proteolytic activities in phosphoric acid-etched dentine by etch-and-rinse adhesives *Biomaterials* **27** 4470–6

[127] De Munck J, Mine A, Van den Steen P E, Van Landuyt K L, Poitevin A, Opdenakker G and Van Meerbeek B 2010 Enzymatic degradation of adhesive–dentin interfaces produced by mild self-etch adhesives *Eur. J. Oral Sci.* **118** 494–501
[128] Osorio R, Yamauti M, Osorio E, Ruiz-Requena M E, Pashley D H, Tay F R and Toledano M 2011 Zinc reduces collagen degradation in demineralized human dentin explants *J. Dent.* **39** 148–53
[129] Sauro S, Osorio R, Watson T F and Toledano M 2012 Therapeutic effects of novel resin bonding systems containing bioactive glasses on mineral-depleted areas within the bonded-dentine interface *J. Mater. Sci., Mater. Med.* **23** 1521–32
[130] de Morais R C, Silveira R E, Chinelatti M A and Pires-de F D 2016 Biosilicate as a dentin pretreatment for total-etch and self-etch adhesives: *in vitro* study *Int. J. Adhes. Adhes.* **70** 271–6
[131] Tirapelli C, Panzeri H, Lara E H, Soares R G, Peitl O and Zanotto E D 2011 The effect of a novel crystallised bioactive glass-ceramic powder on dentine hypersensitivity: a long-term clinical study *J. Oral Rehab.* **38** 253–62
[132] de Morais R C, Silveira R E, Chinelatti M, Geraldeli S and de Carvalho Panzeri Pires-de-Souza F 2018 Bond strength of adhesive systems to sound and demineralized dentin treated with bioactive glass ceramic suspension *Clin. Oral Invest.* **22** 1923–31
[133] Carneiro K K, Araujo T P, Carvalho E M, Meier M M, Tanaka A, Carvalho C N and Bauer J 2018 Bioactivity and properties of an adhesive system functionalized with an experimental niobium-based glass *J. Mech. Behav. Biomed. Mater.* **78** 188–95
[134] Kim Y M, Kim D H, Song C W, Yoon S Y, Kim S Y, Na H S, Chung J, Kim Y I and Kwon Y H 2018 Antibacterial and remineralization effects of orthodontic bonding agents containing bioactive glass *Korean J. Orthod.* **48** 163–71
[135] Zafar M S, Amin F, Fareed M A, Ghabbani H, Riaz S, Khurshid Z and Kumar N 2020 Biomimetic aspects of restorative dentistry biomaterials *Biomimetics* **5** 34
[136] Ferracane J L 2017 Models of caries formation around dental composite restorations *J. Dent. Res.* **96** 364–71
[137] Esteves C M, Ota-Tsuzuki C, Reis A F and Rodrigues J A 2010 Antibacterial activity of various self-etching adhesive systems against oral streptococci *Oper. Dent.* **35** 448–53
[138] Profeta A C 2014 Dentine bonding agents comprising calcium-silicates to support proactive dental care: origins, development and future *Dent. Mater. J.* **33** 443–52
[139] Khvostenko D, Mitchell J C, Hilton T J, Ferracane J L and Kruzic J J 2013 Mechanical performance of novel bioactive glass containing dental restorative composites *Dent. Mater.* **29** 1139–48
[140] Salehi S, Gwinner F, Mitchell J C, Pfeifer C and Ferracane J L 2015 Cytotoxicity of resin composites containing bioactive glass fillers *Dent. Mater.* **31** 195–203
[141] Chatzistavrou X, Velamakanni S, DiRenzo K, Lefkelidou A, Fenno J C, Kasuga T, Boccaccini A R and Papagerakis P 2015 Designing dental composites with bioactive and bactericidal properties *Mater. Sci. Eng.* C **52** 267–72
[142] Chatzistavrou X, Lefkelidou A, Papadopoulou L, Pavlidou E, Paraskevopoulos K M, Fenno J C, Flannagan S, González-Cabezas C, Kotsanos N and Papagerakis P 2018 Bactericidal and bioactive dental composites *Front. Physiol.* **9** 103
[143] Tarle Z, Hickel R and Ilie N 2018 Dentin bond strength of experimental composites containing bioactive glass: changes during aging for up to 1 year *J. Adhes. Dent.* **20** 4

[144] Par M, Tarle Z, Hickel R and Ilie N 2019 Mechanical properties of experimental composites containing bioactive glass after artificial aging in water and ethanol *Clin. Oral Invest.* **23** 2733–41

[145] Yang S Y, Piao Y Z, Kim S M, Lee Y K, Kim K N and Kim K M 2013 Acid neutralizing, mechanical and physical properties of pit and fissure sealants containing melt-derived 45S5 bioactive glass *Dent. Mater.* **29** 1228–35

[146] Nicolae L C, Shelton R M, Cooper P R, Martin R A and Palin W M 2014 The effect of UDMA/TEGDMA mixtures and bioglass incorporation on the mechanical and physical properties of resin and resin-based composite materials *Conf. Papers Sci.* **2014** 646143

[147] Oral O, Lassila L V, Kumbuloglu O and Vallittu P K 2014 Bioactive glass particulate filler composite: effect of coupling of fillers and filler loading on some physical properties *Dent. Mater.* **30** 570–7

[148] Yli-Urpo H, Vallittu P K, Närhi T O, Forsback A P and Väkiparta M 2004 Release of silica, calcium, phosphorus, and fluoride from glass ionomer cement containing bioactive glass *J. Biomater. Appl.* **19** 5–20

[149] Kandaswamy D, Rajan K J, Venkateshbabu N and Porkodi I 2012 Shear bond strength evaluation of resin composite bonded to glass-ionomer cement using self-etching bonding agents with different pH: *in vitro* study *J. Conserv. Dent.: JCD* **15** 27

[150] Matsuya S, Matsuya Y and Ohta M 1999 Structure of bioactive glass and its application to glass ionomer cement *Dental Mater. J.* **18** 155–66

[151] Khoroushi M, Mousavinasab S M, Keshani F and Hashemi S 2013 Effect of resin-modified glass ionomer containing bioactive glass on the flexural strength and morphology of demineralized dentin *Oper. Dent.* **38** E21–30

[152] Yli-Urpo H, Närhi M and Närhi T 2005 Compound changes and tooth mineralization effects of glass ionomer cements containing bioactive glass (S53P4), an *in vivo* study *Biomaterials* **26** 5934–41

[153] Yli-Urpo H, Närhi T and Söderling E 2003 Antimicrobial effects of glass ionomer cements containing bioactive glass (S53P4) on oral micro-organisms *in vitro* *Acta Odontol. Scand.* **61** 241–6

[154] Valanezhad A, Odatsu T, Udoh K, Shiraishi T, Sawase T and Watanabe I 2016 Modification of resin modified glass ionomer cement by addition of bioactive glass nanoparticles *J. Mater. Sci., Mater. Med.* **27** 9

[155] Prabhakar A R and Basappa N 2010 Comparative evaluation of the remineralizing effects and surface micro hardness of glass ionomer cements containing bioactive glass (S53P4): an *in vitro* study *Int. J. Clin. Pediatr. Dent.* **3** 69

[156] Brannstrom M, Johnson G and Nordenvall K J 1979 Transmission and control of dentinal pain: resin impregnation for the desensitization of dentin *J. Am. Dent. Assoc.* **99** 612–8

[157] West N X, Lussi A, Seong J and Hellwig E 2013 Dentin hypersensitivity: pain mechanisms and aetiology of exposed cervical dentin *Clin. Oral Invest.* **17** 9–19

[158] Orchardson R and Gillam D G 2000 The efficacy of potassium salts as agents for treating dentin hypersensitivity *J. Orofac. Pain* **14** 9–19

[159] Absi E G, Addy M and Adams D 1987 Dentine hypersensitivity: a study of the patency of dentinal tubules in sensitive and non-sensitive cervical dentine *J. Clin. Periodontol.* **14** 280–4

[160] Gillam D G 1997 Clinical trial designs for testing of products for dentine hypersensitivity—a review *J. West. Soc. Periodontol./Periodont. Abst.* **45** 37–46

[161] Pradeep A R and Sharma A 2010 Comparison of clinical efficacy of a dentifrice containing calcium sodium phosphosilicate to a dentifrice containing potassium nitrate and to a placebo on dentinal hypersensitivity: a randomized clinical trial *J. Periodontol.* **81** 1167–73
[162] Montazerian M and Zanotto E D 2017 A guided walk through Larry Hench's monumental discoveries *J. Mater. Sci.* **52** 8695–732
[163] Gendreau L, Barlow A P and Mason S C 2011 Overview of the clinical evidence for the use of NovaMin in providing relief from the pain of dentin hypersensitivity *J. Clin. Dent.* **22** 90–5
[164] Ashwini S, Swatika K and Kamala D N 2018 Comparative evaluation of desensitizing efficacy of dentifrice containing 5% fluoro calcium phosphosilicate versus 5% calcium sodium phosphosilicate: a randomized controlled clinical trial *Contemp. Clin. Dent.* **9** 330
[165] Brauer D S, Karpukhina N, O'Donnell M D, Law R V and Hill R G 2010 Fluoride-containing bioactive glasses: effect of glass design and structure on degradation, pH and apatite formation in simulated body fluid *Acta Biomater.* **6** 3275–82

IOP Publishing

Glass-based Materials
Advances in energy, environment and health
Sathish-Kumar Kamaraj and Arun Thirumurugan

Chapter 5

Fiber optic sensors for environmental applications

Daniel A May-Arrioja, Natanael Cuando-Espitia, Amado M Velázquez-Benítez and Juan Hernández-Cordero

Environmental monitoring is important because it provides critical information to protect not only the environment but also because it can help to prevent hazards for human life. In addition to natural disasters that can occur without human intervention, the growth in human settlements and all their associated activities have significantly impacted the environment and human health. As the human population keeps increasing, the need for living spaces causes cities to grow sometimes without planning, and settlements near areas prone to disasters have increased accordingly. This is directly related to never-ending industrial development and energy usage, which has significantly increased the pollution of air, water, and soil. In fact, the significant increase in greenhouse gases and its direct impact on the Earth's temperature is related to extreme weather conditions, food supply disruptions, and increased wildfires [1–4]. Therefore, constant monitoring and eventual prediction of disasters or hazardous conditions for humans is extremely important. Over the years, different sensors and sensor systems have been developed using a wide variety of approaches. Nevertheless, in some applications, the ambient conditions are extreme, and sensors can either fail to withstand or cannot properly operate under such conditions. In some cases, sensors and systems are prone to damage, increasing maintenance and equipment replacement costs over short periods. This scenario has fueled the development and applications of novel technologies that can overcome such limitations and has further driven the improvement of the performance of the sensors. Fiber optic sensors exhibit many exciting features that make them ideal for harsh environments and enhance the sensing capabilities and, in some cases, without the need to invest in new infrastructure. This chapter provides insight into the basic concepts behind fiber optic sensors and systems while overviewing different environmental monitoring and detection applications.

doi:10.1088/978-0-7503-5904-7ch5

5.1 Fiber optics sensing technology

Fiber optic sensors send light (continuous wave or pulsed) along an optical fiber and measure either the transmitted or reflected light. Light traveling along the optical fiber can be modulated by a physical variable, leading to intensity, wavelength, phase, or polarization changes. By measuring such changes, detecting and monitoring variations of the physical variable of interest is possible. High sensitivity and precision can be achieved since measurements are carried out in the optical fiber. Fiber optic sensors are also compact, lightweight, and immune to electromagnetic interference. Fiber optic sensor technology further provides the capability of multiplexing many sensors, and sensors can also operate in harsh environments without compromising their performance. Advances in optical fiber manufacturing and the advent of fiber processing techniques and equipment have led to the development of novel optical fiber structures, which have led to the development of novel fiber optic sensors. All these features make this sensing technology superior to traditional sensors and have fueled their application in different industries as diverse as healthcare, biosensing, aerospace, geophysics, infrastructure, oil and gas, and environmental monitoring, to mention a few. This section briefly describes the basic operation mechanism of fiber optic sensors and the different operating environments where they can be applied.

5.1.1 Basic sensing mechanisms

The field of fiber optic sensors has made significant advancements over the past 30 years. Although various approaches and techniques have been utilized for designing and building sensing systems, two categories of fiber sensors can be identified, and these are referred to as intrinsic and extrinsic sensors [5]. In brief, intrinsic fiber sensors are created to convert the environmental parameter being measured into a change in the characteristics of the guided wave by altering the material properties of the fiber itself. A schematic representation of the intrinsic sensing mechanism is presented in figure 5.1(a). Some physical parameters typically measured using intrinsic sensing include temperature, stress, and vibration. In these sensors, the environmental parameter modulates a feature of the guided wave, namely, intensity, polarization, phase, or spectral content. Intrinsic sensors are fabricated to optimize the interaction between the environment and the optical fiber by, for example, exposing the silica cladding to the target stimuli.

In extrinsic fiber sensors, the light guided by the fiber interacts directly with the targeted environmental parameter. The guided wave thus leaves the optical fiber and can be either recaptured by reflection into the same fiber or transmitted to another fiber. In some cases, additional materials can be used to provide adequate sensitivity to specific environmental parameters. Figure 5.1(b) shows schematically the mechanism for extrinsic fiber sensing. The basic idea is to allow the guided light to interact with the sensing material and transport the modulated signal through the optical fiber. This strategy opens the door to a different number of parameters to be detected as the ability to react to an external perturbation is given by a material that can be engineered and fabricated for a particular application. Detection of pH in

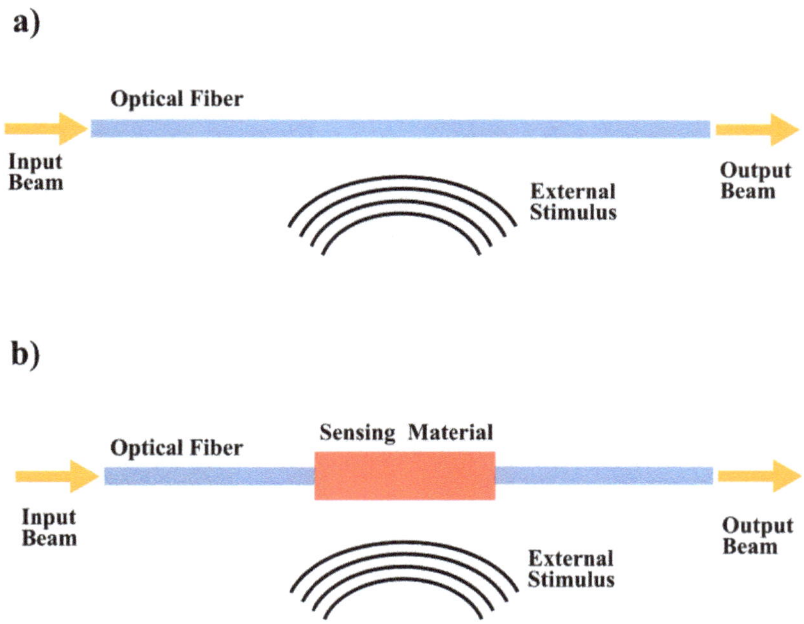

Figure 5.1. Schematic representation of intrinsic (a) and extrinsic (b) fiber optic sensing strategies.

water samples or extremely small amounts of hazardous gasses in the air are two examples of environmental parameters that can be measured using extrinsic sensors. Moreover, as silica may effectively serve as substrate material, fiber optics can readily be functionalized with bio-responsive materials, leading to the realization of immunofluorescence techniques using optical fibers.

In general, extrinsic fiber optic sensors can exhibit improved specificity and sensitivity compared to intrinsic sensors. However, they typically require more complex fabrication techniques and can be less cost-effective. Intrinsic fiber optic sensors are generally more straightforward to fabricate as they do not require additional materials. An important consequence of the operation of intrinsic sensors without any additional material is the possibility of implementing intrinsic sensing techniques using the existing fiber cable infrastructure. This characteristic has been instrumental in performing measurements of temperature and pressure of seawater using existing communication submarine cables. To overcome the fiber optic material's limited sensitivity, intrinsic sensors use more complex readout methodologies, which may be seen as a disadvantage compared to extrinsic sensors.

5.1.2 Point and multi-point sensors

Point-type fiber optic sensors are designed to acquire measurements within a few hundred microns to some centimeters in length along the optical fiber. Since the optical fiber is originally designed to be insensitive to external changes, a sensitive structure must be fabricated either by directly processing or splicing a fiber device at a specific position along the optical fiber. The simplest way to detect external

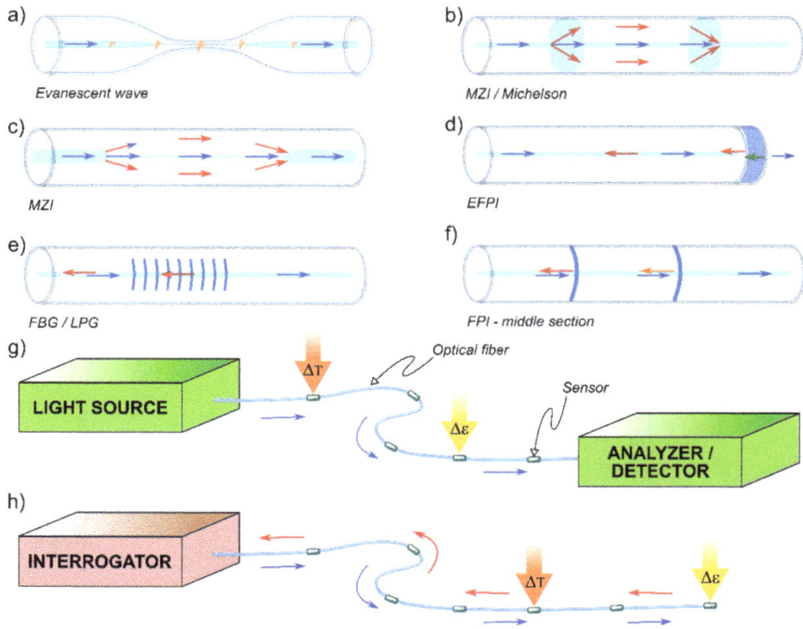

Figure 5.2. Several point-type sensors can be located along a single optical fiber, such as (a) tapered fiber, (b) MZI/Michelson, (c) MZI, (d) extrinsic FPI, (e) FBG, and (f) intrinsic FPI. The sensors are wavelength-encoded, and multiple sensors can be spectrally interrogated by an optical spectrum analyzer or a tunable laser and a photodetector. Schematic representations of two possible quasi-distributed (multi-point) fiber optic sensing systems operating in (g) transmission and (h) reflection. Variations in strain ($\Delta\varepsilon$) and temperature (ΔT) are some parameters that can be monitored with these arrangements.

perturbations is to reduce the fiber diameter by tapering a small fiber section, as shown in figure 5.2(a). This effectively reduces the fiber core to a few microns and, in some cases, to a few hundred nanometers, thus allowing the propagating mode to expand beyond the optical fiber cladding enhancing the detection of perturbations via the evanescent field [6]. A different approach is to splice small sections of different optical fibers to obtain interferometric structures. For instance, a Mach–Zehnder interferometer (MZI) is fabricated by splicing a small section of multimode fiber (MMF) to a single-mode fiber (SMF) to couple light into both the core and cladding of the SMF. After light propagates a small length of the SMF, another MMF section with a similar length is spliced to recouple the light from the cladding to the SMF core, see figure 5.2(b). This recoupling generates interference between light from the core and from the cladding, and thus a sinusoidal modulation is observed when a wide spectral source is transmitted. The evanescent field of the light traveling in the cladding of the SMF can be affected by external perturbations, which modify the original interference pattern. A Michelson interferometer is obtained if this structure is cleaved exactly in the middle and the reflected interference pattern is monitored. An MZI can also be obtained by splicing a section of a fiber with a small core between fibers with a larger core diameter. The mode mismatch will couple light into the cladding, as shown in figure 5.2(c), which will then be recoupled after propagating along the small core section, effectively producing an interference pattern. Many other structures for MZIs have been developed over the years, always aiming at having light traveling along the fiber cladding [7].

Another fiber device used as a point sensor is the Fabry–Pérot interferometer (FPI). This type of interferometer can be easily fabricated by attaching a thin layer at the tip of an SMF. For example, this has been achieved using a section of MMF and dipping the tip in a polymer. In this configuration, as shown in figure 5.2(d), light will experience a first reflection at the glass–glass or glass–polymer interface (red arrow) and a second at the glass–air or polymer–air interface (green arrow). The reflected light can be used to monitor the interference pattern generated in the device when using a wide spectral source to interrogate the sensor. If the optical properties of the material contained within the reflective surfaces (e.g., a polymer or air) are affected by a physical parameter, the interference pattern will experience either a phase shift or a contrast reduction, thus allowing for monitoring variations of the physical parameter.

The most successful intrinsic fiber device in terms of commercial applications is arguably the fiber Bragg grating (FBG). An FBG is fabricated by 'imprinting' a periodic refractive index modulation in a short fiber length, see figure 5.2(e). These periodic variations can be produced using different techniques such as holography, phase masks, and point-by-point modifications of the glass' refractive index [8]. This periodic modulation provides multiple reflections that add in-phase and ultimately reflect a very narrow range of wavelengths (typically less than 0.1 nm) and can thus be interrogated with a broadband source or a tunable laser. The peak wavelength of these devices can be set during the fabrication process by selecting the proper period for the refractive index modulation. When the FBG is heated or strained, the effective refractive index and period of the FBG are modified, and the peak wavelength shifts accordingly. This feature allows for measuring strain and temperature using FBGs [9]. Another popular intrinsic fiber sensor for measuring strain and temperature variations is the FPI, which can be obtained upon incorporating two reflectors inside the SMF, as shown in figure 5.2(f). These mirrors can be fabricated using different techniques such as micromachining, FBGs, chemical etching, and thin film deposition [7].

Most point-type sensors can be used in transmission, using a light source in one of the fiber ends and an analyzer or a detector at the opposite end, as shown in figure 5.2(g). FPIs fabricated on the fiber tip and FBG sensors can also be used in reflection mode, allowing one to allocate the source and the analyzer/detector at the same fiber end (see figure 5.2(h)). As discussed above, an important feature of these sensors is their spectral response. Since they are wavelength-encoded, each sensor has a characteristic spectral response, thus allowing for the realization of wavelength multiplexed configurations. This has provided a means to perform multi-point measurements (i.e., quasi-distributed) along a single optical fiber. In principle, most point-type sensors can be used for these quasi-distributed sensing systems, including tapered sensors that can be fabricated to exhibit a sinusoidal spectral response. FBGs and FPIs are mostly used for quasi-distributed sensing systems operating in reflection [10, 11] while the rest are most commonly used in sensing systems operating in transmission configurations.

5.1.3 Distributed fiber sensors

Distributed sensing refers to spatially resolved measurements, and optical fibers are arguably the only means to achieve such functionality in sensing systems. Although light propagates practically without losses along an optical fiber, photons interact with the glass structure leading to scattering processes during propagation. Since external parameters such as temperature and strain can perturb the fiber structure, scattering events along the fiber can be used for sensing. Scattering-based sensors rely mainly on monitoring three kinds of events: Rayleigh, Brillouin, or Raman scattering [12]. The strategy to use these as sensing mechanisms is based on monitoring the intensity and the wavelength (frequency) of the backscattered photons along the optical fiber. As illustrated in figure 5.3, each scattering event leads to a backscattered signal with a specific spectral location and amplitude. When the fiber is exposed to changes in temperature or under different strain conditions, these spectral features will change accordingly.

Rayleigh scattering is known as elastic scattering because the backscattered signal is at the same wavelength as that of the propagating light. In contrast, Raman and Brillouin scattering are referred to as inelastic events because the backscattered photons show a shift in wavelength owing to their interactions with acoustic phonons (Brillouin) and molecular vibrations of the silica lattice [12]. The shift in wavelength due to these inelastic scattering effects can be either toward shorter wavelengths (anti-Stokes spectral components) or longer wavelengths (Stokes spectral components). Because all three scattering events are different in nature, the backscattered signal arising from each of them is located at different

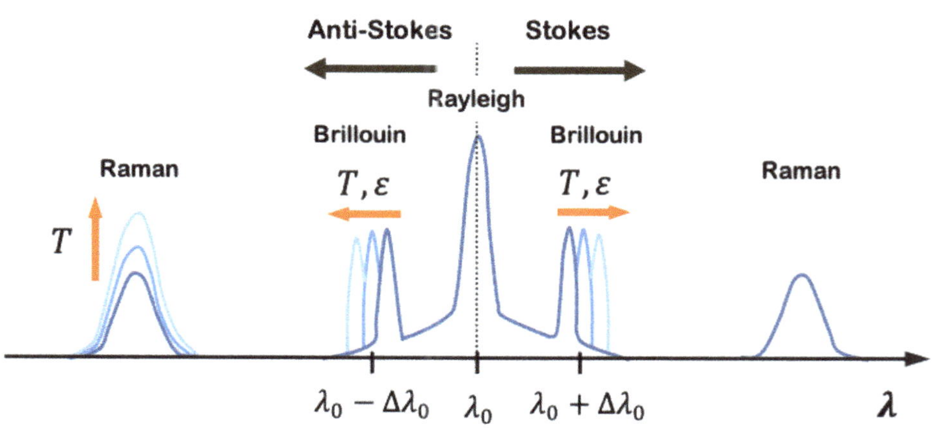

Figure 5.3. Spectral features of the backscattered light used for distributed sensing: while Rayleigh scattering occurs at the same wavelength of the propagating light, Brillouin and Raman scattering show spectral components to the right (Stokes) and the left (anti-Stokes) of the central wavelength. The spectral features (amplitude or peak wavelength) of Brillouin and Raman signals are affected by temperature and strain on the fiber.

wavelengths, and they are also sensitive to different parameters. Brillouin scattering is affected by both temperature and strain; in contrast, Raman scattering is only affected by temperature. Furthermore, as illustrated in figure 5.3, temperature changes modify the amplitude of the anti-Stokes Raman signal but lead to a shift in wavelength in the Brillouin spectrum. These features allow for obtaining information about both parameters simultaneously despite the potential cross-sensitivity of the Brillouin effect. Meanwhile, Rayleigh scattering allows for measuring temperature, pressure, and acoustic signals, depending on the interrogation scheme used for monitoring the backscattered light [12, 13].

Optical reflectometry techniques enable distributed sensing capabilities that are attainable through scattering effects. For sensing applications, reflectometry can be performed either in the time domain (optical time-domain reflectometry, OTDR) or the frequency domain (optical frequency-domain reflectometry, OFDR). OTDR was first introduced to assess losses along fiber optic links, as attenuation is measured through time-of-flight measurements. A short light pulse (typically tens of nanoseconds) is launched into the sensing fiber and the Rayleigh backscattered signal is detected as a function time using a fast photodetector synchronized with the pulsed source (see figure 5.4). The factors affecting the spatial resolution, sensitivity, and signal-to-noise (SNR) ratio in OTDR measurements include the pulse duration, the optical power, and the fiber characteristics (numerical aperture, refractive index profile, and mode field diameter). Notice that these factors are inherent to the measuring technique; the sensitivity to the parameter to be measured (e.g., temperature, pressure, stress, etc) will depend on the features of the fiber used for sensing, as described in subsequent sections. Variations of this technique include phase-sensitive and polarization-sensitive OTDR, in which either the phase or polarization of the backscattered signal is used for monitoring disturbances along the fiber [12, 13]. These schemes are particularly useful for sensors based on Rayleigh backscattering.

In OFDR, the intensity of the laser remains nominally constant, but the wavelength is swept linearly with time (see figure 5.4). Light is split into two signals: one is used as a reference, and the other is launched into the fiber as a probe beam.

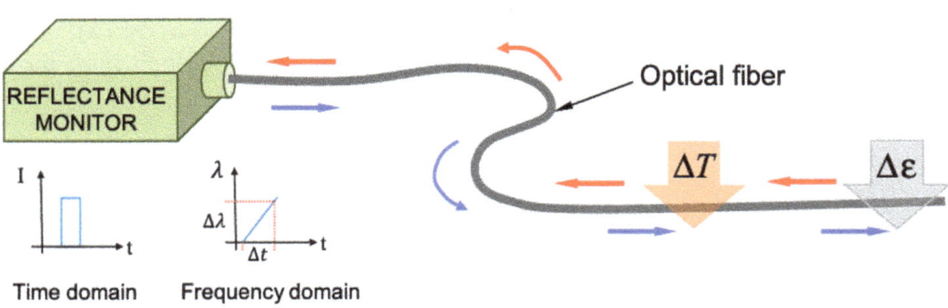

Figure 5.4. Schematic representation of optical reflectometry techniques for scattering-based distributed sensing. The reflectance monitor system includes a modulator and a fast detector synchronized with a laser source. A pulsed laser is used for time-domain measurements (OTDR), and a tunable laser is used for frequency-domain measurements (OFDR). Time-of-flight measurements allow for locating induced perturbations (due to strain ($\Delta\varepsilon$) or temperature (ΔT)) along the optical fiber.

Because the backscattered signal is essentially a time-delayed version of the reference beam, both can be compared coherently, and information about perturbations along the fiber can be readily obtained [12, 14, 15]. OFDR offers improved spatial resolution and dynamic range compared to OTDR, albeit requiring more elaborated modulation and detection schemes. The spatial resolution is inversely proportional to the span of frequencies covered with the laser. As long as a linear relation between the wavelength and the modulation frequency can be sustained, the information from the backscattered signal can be retrieved coherently. As in any other sensing system, an increase in spatial resolution reduces the measurement range; hence, OFDR systems are typically limited to ranges of less than tens of meters. However, this can be increased if lasers with long coherent lengths are used in the system or by means of more sophisticated frequency modulation schemes [12]. Finally, we mention optical low-coherence reflectometry (OLCR), an alternative to OTDR and OFDR [12]. This is essentially an interferometric technique in which the optical fiber is placed in one arm of a Michelson interferometer, and a moving mirror is used in the reference arm. Using a broadband light source, the backscattered light from the fiber is analyzed, and the location along the fiber of the reflected signal can be obtained from the position of the mirror. OLCR measurements can provide micron-level spatial resolutions, but the measurement ranges are less than a few meters. Since most environmental sensing applications typically seek long measurement ranges, OTDR and OFDR are the most commonly used techniques.

5.2 Land monitoring and detection

Natural land hazards can occur without any warning and can have a significant impact on infrastructure and human lives. In some cases, danger comes from human-made situations, such as people living near hazardous areas or influencing the trigger of a land hazard. Regardless of its source, we must be aware that we live in a constantly changing land where hazardous life-threatening events can occur, and we must be prepared or at least have some time to react. Events such as earthquakes, volcanic eruptions, landslides, and sinkholes, in many cases, are difficult to predict. Nevertheless, having systems that can help to issue warnings or detect when an event occurs is highly critical. In this section, we explore the application of fiber optic sensors for monitoring and detecting such land hazard events.

5.2.1 Earthquakes

An earthquake is a natural event that can significantly impact cities even when they are not close to the earthquake epicenter. The main hazard directly related to earthquakes is the effect of ground shaking, which can have profound consequences for buildings, infrastructure, and human life. When an earthquake occurs under the ocean, we also must worry about tsunamis because they can cause considerable damage to the coastline. Currently, it is impossible to predict when an earthquake will occur. Therefore, researchers have been developing instruments to monitor and, hopefully, predict earthquakes one day. Among the different physical features to monitor an earthquake, seismic waves have demonstrated to be very effective [16]. Scientists have long used seismometers to measure seismic waves to monitor and try to predict earthquakes. Over the years, a network of seismometers has been installed around the globe, providing important data for research and monitoring. The main drawbacks of seismometers are related to their high installation and maintenance costs and their limited number, counting 152 currently installed worldwide, which provides limited spatial coverage. The issues only worsen when dealing with offshore areas due to the extreme conditions for underwater operations. Therefore, novel techniques are required to cover the gaps observed with seismometers.

Recently, a technique known as distributed acoustic sensing (DAS) has been used to monitor and detect earthquakes [17, 18]. In a DAS system, coherent pulses of light are sent along the optical fiber, and backscattering occurs due to imperfections. When seismic waves disturb the optical fiber, the induced strain modifies this backscattered light, and the locations and magnitudes of the perturbations along the cable can be obtained using signal processing techniques (see the distributed sensing section for further details). In essence, these ‘imperfections’ act as distributed sensors along the length of the fiber cable. Experimental results demonstrate that DAS systems have broad bandwidth and can record seismic waves with high fidelity. As shown in figure 5.5, DAS systems can detect seismic waves, and the resulting signals are quite similar to those obtained from a traditional seismometer. A key advantage of DAS technology is that it can be deployed in dedicated optical fiber cables or using existing telecommunications fiber cables that are not currently in use for data transmission (the so-called dark fibers). This also means that offshore and

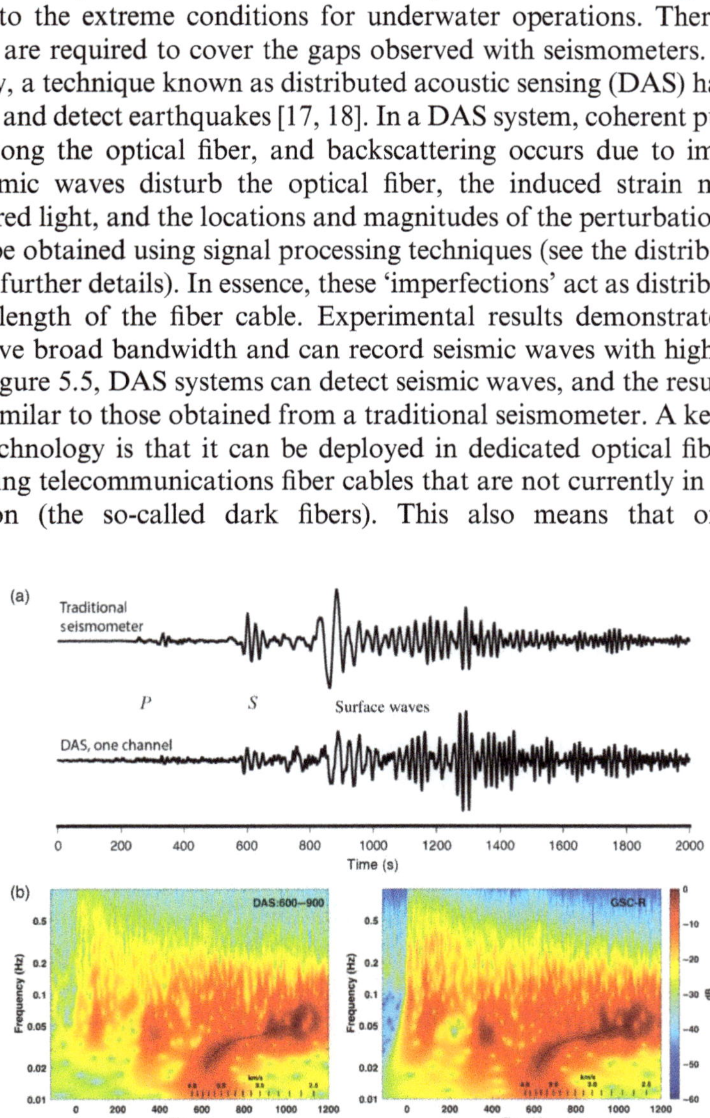

Figure 5.5. Comparison between broadband DAS and broadband seismic station (CI.GSC) for the 2018 Mw 7.5 Honduras earthquake. (a) Waveform of one DAS channel and radial-component of the CI.GSC. (b) Spectrograms from stacked DAS waveforms and GSC radial components. Reproduced with permission from [17].

transoceanic fiber optic cables laid on the ocean floor can be used as seismometers to detect earthquakes under the sea. These advantages are extremely important because it means that earthquakes can be monitored and detected without requiring either the installation of new sensing instruments or additional maintenance of the existing infrastructure, thus providing a cost-effective solution.

DAS systems have been employed in the last few years to monitor seismic waves in different scenarios. In a recent report, dedicated on-land optical fibers in three different locations were used to record seismic information using DAS systems. A comparison of DAS results with those recorded from a standard seismometer showed an excellent agreement between the amplitude and phase features of the recorded signals [19]. In 2020, the vibrations generated during the Rose Parade in Pasadena, California, were monitored using a DAS system. Using a telecom optical fiber cable located underneath the Rose Parade's route, different seismic signatures allowed the researchers to detect the movement and vibrations caused by different elements of the parade: the police motorcycle squad, floats, and marching bands were readily detected. In fact, the system provided enough information to award two new prizes to the heaviest float and the loudest band, which were easily determined based on the readings from the DAS system [20]. Recent studies have also demonstrated a DAS system for monitoring seismic waves from an offshore fiber cable in Chile. As mentioned earlier, earthquakes in the ocean are important to detect because they can trigger tsunamis [21]. Aside from providing a similar response to that from a seismometer, the DAS system detected the P-wave arrival 25 s before the on-land seismometer. This time window can be critical for alerting coastal areas and demonstrates the advantage of DAS systems for earthquake early warning (EEW) systems [22].

Although DAS systems offer an excellent way to retrieve seismic information from undersea fiber cables, researchers are also exploring new techniques. In a recent experiment, microwave frequencies were transmitted back and forth along an optical fiber to assemble a sensitive microwave interferometer for detecting earthquakes. The key advantage of this approach is that high spectral purity microwave oscillators (with sub-Hz frequency stability) are more affordable than their laser counterparts. The seismic signals acquired with the interferometer exhibit a remarkable agreement with those obtained from accelerometers and DAS systems, potentially impacting cost reduction [23]. An interesting approach was recently reported based on active phase noise cancellation (PNC). PNC typically generates optical phase change measurements, which allow frequency stabilization in metrological fiber networks. These phase changes can be correlated with ground deformation and vibrations, which provides a distributed seismic sensor. Validation through spectral-element wavefield simulations and real data from an earthquake, displaying a match in phase and amplitude, showed that the system can provide quantitative earthquake detection and characterization. Since PNC systems are compatible with inline amplification, this technique has the potential to operate beyond 1000 km, making it ideal for transoceanic EEW [24]. In another approach, the state of polarization (SOP) of light signals transmitted through the fiber optic cables is continuously monitored. When seismic or water waves press or shake the cable, the perturbations along the fiber introduce changes in the SOP that can be detected and analyzed. Impressively, routine monitoring of standard optical telecommunication channels enabled detecting vibrations with peak displacement as low as 0.1 mm along a 10 000-km submarine cable or over round-trip channels spanning 20 000 km. The key advantages of this technique are that it can use equipment that is already installed in telecom systems, and in addition to earthquakes, it can also detect tsunamis [25].

Advances in machine learning (ML) are also being applied to analyze large amounts of data recorded by DAS systems. In recent work, seismic waveforms from standard seismometers were used to train neural networks to identify critical earthquake features. After training, the neural networks could reliably detect seismic waves from DAS recordings. Regardless of the differences between waveforms from DAS and seismometers, the trained models exhibited accuracies between 93.86% and 96.94% [26]. At the end of last year, an ML algorithm was trained with five years of seismic data from 2016 to 2020, particularly emphasizing seismic waveforms that occur a week before an earthquake. The algorithm was then used to evaluate real-time data acquired from seismometers during the year 2021, aiming to predict earthquakes the week after analyzing the data of the current week. Although the algorithm was able to predict 14 earthquakes, it also forecasted eight that did not occur and missed one [27].

5.2.2 Volcanic eruptions

Volcanic eruptions are natural disasters that can be extremely hazardous for cities close to the volcano. The situation worsens when the volcano is inactive since the danger of a potential eruption is underestimated. Monitoring volcanic activity becomes important as a warning system and a means to understand the diversity of underground processes that eventually trigger a volcanic eruption. Different variables can be measured for volcano monitoring, including earthquakes and tremors, ground motion, gas composition, and changes in local electric and magnetic fields. Monitoring ground motion and earthquakes has been the most effective way to predict a volcanic eruption. The main issue with continuous monitoring of volcanoes is the harsh environment that can easily damage installed sensors near the volcano. In this context, DAS systems have been implemented to monitor volcanoes because they can effectively monitor earthquakes, and the installed fiber underground is not significantly affected by ashes and rocks near the volcano.

Dedicated optical fibers have been installed to acquire information with DAS systems at the Etna volcano (Italy). A comparison of DAS data with a seismic array showed good agreement in general, thus validating the DAS measurements. The information provided by DAS systems allows for investigating different complex subsurface processes while acquiring data at a safe distance. Additionally, since DAS acquires information from many channels, it provides information from local perturbations, fault zones, as well as explosions and their location [28, 29]. In recent experiments, DAS systems installed in the Azuma volcano (Japan) and the

Stromboli volcano (Italy) allowed researchers to locate the sources of seismic waves. Remarkably, they were able to locate the seismic sources regardless of the high levels of strain noise [30, 31]. A telecom fiber optic cable connecting Volcano Island to Sicily (Italy) with both on-shore and offshore sections was also used to monitor the volcano. Results from these experiments showed that even when the fiber cable is not optimally coupled to the ground, it is indeed possible to detect low-frequency seismic waves. By combining conventional and ML algorithms, automatic detection of seismic events, including feature extraction and event classification, has also been demonstrated [32]. Monitoring of volcanoes under ice was tested at the Grímsvötn volcano using a dedicated fiber optic cable installed 30 cm under the ice. This arrangement provided good-quality data and allowed the detection of small deformations of a few nanostrain/s at frequencies ranging from $\sim$0.1 to 20 Hz. Placing the fiber cable in the ice cap proved beneficial since it effectively amplified volcanic tremors, which could not be detected under normal circumstances [33, 34].

5.2.3 Landslides

Several factors are involved in the development of cracks leading to surface fragmentation occurring on the Earth's surface, particularly on slopes. Internally, factors such as the gravitational force within rock and soil masses and groundwater play an important role. Externally, earthquakes and rainfall are also recognized as relevant aspects leading to surface fragmentation. Rainwater subsequently infiltrates these cracks, creating sliding surfaces within the slope, causing the rock and soil mass to slide and resulting in landslide disasters. To prevent disasters of this kind, early detection of soil movement can crucially provide enough time for evacuation. Landslides have been monitored using different techniques such as remote sensing (optical, radar, LiDAR), geotechnics (inclinometers, extensometers, tiltmeters, and geophones), geodetic methods (monitoring absolute displacements), hydrogeological (water-related data), and mapping [35]. Recently, fiber optic sensors have been applied to landslide monitoring since they offer several benefits that can contribute to an efficient warning system.

FBGs have been widely used for geotechnical applications, including landslide monitoring [36]. In this application, FBGs are operated as a multi-point sensor to measure strain at many points along an optical fiber. The fiber with many FBGs is placed vertically, and when pushed in their horizontal direction, it works as an inclinometer and can be used to monitor landslides [37–40]. As shown in figure 5.6(a), different points can be selected to insert inclinometers (D1 to D4) in a region with the potential for a landslide to occur. In these points, a borehole is drilled until a stable bedrock layer is reached. As shown in figure 5.6(b), the fiber with the FBG sensors is attached to a tube that is then inserted in the borehole and subsequently filled with cement. After installation and proper calibration, land displacement induces strain in the FBG, which can then be converted to displacement, as shown in figure 5.6(c). This provides an efficient system for monitoring and early warning for landslides.

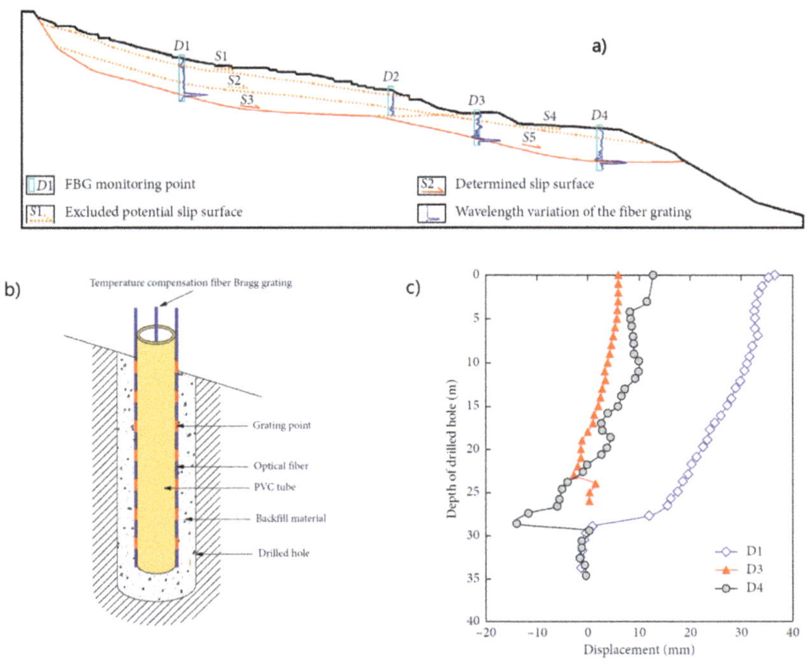

Figure 5.6. (a) Cross-section of the monitoring of internal deformation for the Toudu landslide, (b) schematic diagram of optical fiber installed in the tube and inserted in the borehole, and (c) internal deformation of landslide. Reproduced from [40] CC BY 4.0.

Distributed fiber sensors (DFSs) can also monitor strain along an optical fiber, making them an attractive option for landslide monitoring [12]. Since strain data is acquired along the optical fiber, DFSs can be used in boreholes (similarly to FBG sensors) or lying under a specific area under test. For instance, a polarization-sensitive optical frequency domain reflectometer (P-OFDR) was used in a borehole to perform a field trial in the Sichuan province (China) in 2009. Results showed the potential of this approach as an early warning system with a resolution of 5 cm and a theoretical measuring range of 10 km [41]. OFDR has also been used in a physical model to reproduce a landslide triggered by rainfall, deploying the fiber at a specific depth and with a spatial resolution of 10 mm. Data acquired with the OFDR system was validated upon comparison with information obtained with conventional sensors, which showed excellent agreement [42]. Brillouin Optical Time Domain Analysis (BOTDA) has also been used for landslide monitoring. By installing an optical fiber cable in a road in the mountain resort town of St. Moritz, Switzerland, deformations suffered by the road can help to obtain measurements for landslide prevention [43]. More recently, a BOTDA system was used first in a scale model and later for a real application. A 30 m long optical fiber was attached to the Coroglio Tuff cliff (Italy), and data was studied for three years. As in previous experiments, the results revealed the effectiveness of DFS to detect increases in strain and rock fractures, thus highlighting the excellent potential of DFS for landslide monitoring [44]. Finally, DFS can also be used to monitor and predict the occurrence of sinkholes. Since soil displacement can occur within the sinkhole zone, an optical fiber can be placed in the correct location to assess the strain produced along the fiber [45].

Clearly, DFSs have great potential to be effectively used for real-time magnitude estimation and prediction of seismic waves and ground motion. Nonetheless, they should not be considered the ultimate solution for an EEW system. Instead, this technology should be part of a broader and comprehensive EEW system encompassing various sensors and monitoring technologies.

5.3 Water-related monitoring

Water is indeed essential for humankind. Aside from the essential supply of water on a daily basis for consumption required by humankind, water is also important for energy generation, sanitation, and agriculture, just to name a few critical instances. According to the United Nations, one of the critical aims for humanity in the 21st century is preserving water availability, as freshwater resources suffer from pollutants associated with global changes and globalization [46]. Water contamination from a given application may significantly affect several unrelated areas, increasing the need for specific and dedicated water quality monitoring. In this urgent scenario, fiber optics may represent an instrumental tool for developing robust, versatile, and high-precision sensors for this application. In particular, their small dimensions, biocompatibility, and the fact that silica is a chemically inert material are attractive characteristics of this sensing technology for water quality and related monitoring.

5.3.1 Water quality

Although water bodies are clearly a fundamental component of the environment, water may carry chemicals and pathogens that would turn this asset into a significant health hazard, either in the short or long term. In order to manage and evaluate freshwater quality through a single-number figure of merit, different water quality indices have been proposed [47]. In general, a specific Water Quality Index (WQI) is computed by weighting a set of measurements from a number of predefined chemical, biological, and physical water parameters. Measurements of parameters such as pH, suspended solids, fecal coliforms, and dissolved oxygen are compared to the acceptable range of each parameter to obtain the WQI. Typically, water samples are collected and transported to the laboratory to perform the required testing. In this context, fiber optics offers the possibility of real-time, high-sensitivity, distributed, and dedicated sensing [48].

As in other fiber optic sensors, the basic idea of fiber optic water quality sensing is the modification of a light beam variable (i.e., intensity, phase, or polarization) by means of a specific water parameter. Thus, the value of a given water parameter can be inferred by tracking the changes in the light traveling within the fiber. One of the most straightforward approaches reported for fiber-based water sensing is probably that using unmodified and low-cost optical fiber. This approach further relies on commercially available readout equipment to measure the concentration of ethanol, isopropyl alcohol, and acetone in water [49]. Results with this arrangement using an OFDR sensing scheme show that the swelling of the polymer coating of the low-cost fiber can be readily detected within sub-millimeter resolution. As the solvent diffuses into the polymer coating that covers the fiber, a localized strain is generated, thereby affecting the backscattered light, which is monitored by an interferometric setup. Although this work did not show solvent-specific results, this approach may be modified to include porous-specific materials such as metal–organic frameworks (MOFs), potentially allowing for the development of multi-parameter distributed sensing based on optical fibers.

A similar approach using OTDR and commercial optical fibers has incorporated artificial neural networks to recognize the presence of water [50] detect bacteria and chemical pollutants based on the combination of ML and principal component analysis, as demonstrated using solutions with different concentrations of ethanol in water [51]. Other pollutants can be detected using conventional optical fibers operating in the near-infrared (NIR) region by exposing the evanescent wave to the water medium and tracking the transmission spectra. Kerosene, hexane, palm oil, acetic acid, naphtha, and toluene have been detected with a detection limit of up to 5% using optical fiber tapers [52]. This multi-parameter sensing scheme for monitoring seawater conditions used an array of tapers with different coatings.

Implementing a specific sensing approach in the NIR range (0.7–3 μm) enables the use of various fiber devices and optical equipment originally designed for communication purposes. However, certain water contaminants exhibit significant optical absorption within the mid-infrared spectrum (MIR, 3–30 μm). As a result, optical sensing in the MIR region can provide distinct identifying features ('finger-prints') that enable accurate detection and quantification of particular water pollutants showing high optical absorption. This detection method can be extended to fiber sensing through silver halide fibers with low loss in the MIR range and can effectively transmit the MIR radiation necessary for accurately sensing water contaminants. For instance, parathion, diazinon, and 2,2-dichlorovinyl dimethyl phosphate (DDVP) detection was demonstrated in water solutions with a concentration range of 1–100 ppm [53]. Hydrocarbons such as benzene, toluene, and xylene can also be accurately detected in water using MIR-based fiber sensing techniques [54]. By coating the fiber with a specific sensing material that selectively interacts with the target contaminants, the sensitivity and selectivity of the optical sensor can be enhanced. An example of this approach is the use of ethylene propylene copolymer as a fiber coating material for the detection of volatile organic compounds in groundwater [55]. Halide fiber tapers have also been demonstrated to improve the sensitivity of MIR-based fiber sensors, allowing for increased precision in detecting and quantifying water contaminants. Direct measurement of the transmission spectra from nanoparticle-coated halide fiber taper has also allowed for high-sensitivity detection of xylene [56].

Fiber optics can also effectively transmit light in the visible spectrum and have therefore been used for water quality monitoring assessment in this spectral range. In contrast to NIR applications using fiber devices for telecommunication, fiber systems operating in the visible spectral region benefit from commercially available devices primarily designed for vision-related purposes. Plastic optical fibers are particularly well-suited for visible applications due to their cost-effectiveness and high mechanical flexibility compared to conventional silica optical fibers. Obrovski *et al* developed a sensing probe consisting of a bundle of four plastic fibers ending in a sample container with a mirror at the bottom, positioned in front of the fiber bundle [53]. In this probe, three fibers deliver light from red, green, and blue LEDs to the water sample, while the fourth fiber collects reflected light from the mirror and delivers it to a photodetector. An electronic control synchronizes the emission-detection of each light component, and the analysis of the recorded intensities leads to a differentiation among pollutants such as chlorine, chromium, nitrite, and orthophosphate. Other reported sensors in this context are based on increasing the interaction between the fiber material and the surrounding liquid media by, for example, polishing a section of the plastic fiber to detect perfluorobutanesulfonic acid [57]. Biosensing using fiber optics in the visible spectrum to detect *e.coli* bacterial contamination in water has recently been demonstrated using a modified plastic fiber tip incorporating a double-layer coating of Au and ZnO nanorods [58]. The binding between the *e.coli* and the ZnO nanorods modifies the backscattered signal, thus indicating the presence of the bacteria. In general, sensing schemes based on light scattering and refractive index changes lack specificity, and approaches using more complex chemistry have also been investigated. Silica, the material used in most optical fibers, can be used as a substrate for conventional functionalization methods. For instance, highly selective detection of 17b-estradiol, an endocrine disrupting compound, frequently present in environmental water samples, was accomplished within the nanomolar concentration range by instrumenting a fiber-based fluorescence scheme [59].

5.3.2 Physical parameters of water quality

Water parameters such as pressure, temperature, and salinity offer valuable insights for water monitoring despite not being directly linked to pollution contaminants. Compared to fiber optic sensors for specific pollutants, fiber optic systems for physical parameters have been extensively studied as the sensor's response can readily correlate to the fiber's physical perturbation. Moreover, fiber sensing can be easily applied in water to measure physical parameters due to the small size of fibers and their chemical inertness. In general, these kinds of sensors are based on recording their transmission/reflection spectra and tracking the spectral shift of a predetermined region of maximum/minimum intensity. These spectral features may be obtained using interferometric and periodic fiber devices or photonic guiding effects.

Temperature fiber sensors have exploited sensing elements such as FPI, FBGs, and microfiber resonators [60–62]. In general, these sensors exhibit sensitivities of a few nanometers per °C and up to a few picometers per °C [63]. Different implementations of Fabry–Pérot cavities have been predominantly used on pressure sensing in water media, and research groups have reported sensitivities of 1.071 rad MPa^{-1} using micromachining and 9 nm psi^{-1} using a surface-mountable fiber probe [64, 65]. In contrast to pressure sensors, more diverse approaches have been tested for fiber pH sensing; devices such as long period and Bragg gratings, surface plasmon resonances, MZI, and multimodal interferometers have shown to be useful to detect pH from 1 to 12 [66–70]. This is particularly relevant because pH is an important parameter in other fields, such as biomedical research and industrial processes.

Finally, salinity sensors are also important tools in marine monitoring. Marine salinity affects seawater organisms and provides information on global oceanic currents. Conventional saline sensors rely on the electrical conductivity of seawater [71], but corrosion and electromagnetic interference hinder their performance. Thus, fiber optic sensors are less prone to corrosion and show null electromagnetic interference effects. As in the case of pH sensors, different fiber sensing elements and fundamental mechanisms have been explored in developing salinity fiber sensors to measure saline concentration. Fiber interferometers, fiber gratings, and microfiber resonators are some reported approaches for water salinity measurement [72–74]. Recently, an ultrahigh sensitivity sensor based on a photonic crystal fiber reportedly achieved a sensitivity of 7.5 nm/% [75]. According to the authors, the proposed sensor operates in the whole range of salinity (from 0% to 100%), having structural simplicity and extended longevity.

5.3.3 Heavy metal detection

While small amounts of heavy metals are naturally contained in drinking water and daily food intake, elevated levels of these elements are extremely harmful to human health, potentially leading to conditions such as brain damage, liver issues, and bladder cancer [76]. The rise in naturally existing levels of heavy metals in aquatic environments can be attributed to human actions like mining, farming methods, and industrial waste disposal [77]. Notable examples of common heavy metals found in water include mercury (Hg), arsenic (As), lead (Pb), cadmium (Cd), and chromium (Cr). Like other metals, these substances also exist in different oxidation states due to their coordination and oxidation-reduction properties. Generally, the oxidation states of heavy metals pose more risk than their elemental forms.

Conventional detection methodology of heavy metal ions covers techniques such as spectrophotometry, chemiluminescence, and inductively coupled plasma mass spectrometry. These techniques typically require expensive equipment, sample pretreatment, and provide long response times. In contrast, fiber optic sensing technology offers high-sensitivity detection, faster response times, and *in situ* and distributed measurement capabilities. Due to the rapid growth of optical fiber sensing technologies and functional materials, optical fiber sensors have demonstrated outstanding potential for measuring the concentration of heavy metal ions. Fiber optic heavy metal ion sensors usually integrate a special fiber coating whose optical properties are modified when interacting with a specific metal ion. This, in turn, modifies the features of the light guided by the fiber, and the targeted ion can be detected and, in some cases, quantified. Efforts to develop these types of sensors include absorption-based methods, interferometric sensing, grating-based designs, fluorescence approaches, and plasmonic sensing. Absorption-based sensing has been implemented to detect Hg^{2+} and Cd^{2+} by registering the spectral response of the fiber system and correlating the intensity decay with the concentration of targeted ions [78, 79]. These sensors used dithizone and chitosan as sensitive materials, respectively. Interferometric fiber sensors have been used to detect mainly Ni^{2+}, although Cu^{2+}, Pb^{2+}, and Cd^{2+} detection has also been reported [80, 81]. Chitosan, engineered polymers, and chelating agents have been used as coating materials in interferometric sensors to enhance the interaction of the guided wave with the water environment. Fiber sensors based on fiber gratings have used hydrogels and black phosphorus to detect Cr^{2+} and Pb^{2+} [82, 83]. As before, these sensors rely on modifying the coating layer when in contact with the targeted heavy metal ions. For instance, a volume increase of hydrogel triggered by the incorporation of Cr^{2+} ions can yield a spectral shift via mechanical strain on the fiber cladding [82].

Plasmonic-based sensors represent a highly versatile option for heavy metal detection as the sensitive layer can be designed to exploit the surface plasmon characteristics of metal nanoparticles and antibodies-based coating. Briefly, heavy metal binding to the sensitive layer results in specific spectral changes. Detection of Mn^{2+}, Pb^{2+}, and Cu^{2+} has been achieved using plasmonic-based sensors. In both approaches, the metal-based coating is fabricated to selectively enhance the interaction of the imprinted material with the heavy metal ions [84, 85]. Although plasmonic-based sensors have been shown to provide the lowest limit of detection (0.2×10^{-4} nM [85]), they require precise fabrication to be 'tuned' and exploit the specific plasmonic features needed for detection. Finally, fluorescence-based fiber sensors have also been explored for heavy metal detection. The sensing principle in these kinds of sensors relies on the changes in the fluorescence signal occurring with the bonding of functionalized molecules with heavy metal ions. When functionalized molecules are hybridized with fluorescence materials, the formation of heavy metal complexes may lead to a modification of the fluorescence emission [86, 87]. The heavy metal-induced signal modification can be an increase/decrease in fluorescence intensity or a shift in the spectral emission. Thus, fluorescence-based fiber optic sensors exhibit high specificity. However, the engineered coatings needed for these sensors require elaborated chemical processes and suffer from rapid environmental deterioration.

5.3.4 Sewer disposal

Proper wastewater management is vital for protecting public health, particularly in urban areas. Improper handling of wastewater has the potential to cause substantial harm to the plants and animals within an ecosystem, as well as long-term effects on public health. In addition, unpolluted water sources are becoming increasingly scarce, while the need for drinking water is rising due to population growth and increased industrial activity. This highlights the necessity for improved water treatment facilities with greater capacity as well as the need for sewer infrastructure monitoring. As in other areas, electromechanical and electrochemical sensors have dominated wastewater monitoring technologies. Although these technologies have demonstrated rapid response times in sensing water parameters and a convenient cost efficiency, fiber optic sensors exhibit higher selectivity and high-resolution operation. Moreover, as fiber optic sensors inherit electromagnetic noise immunity and chemical inertness from silica, fiber optic technologies represent an ideal approach for aggressive environments such as water disposal piping.

A particular concern in sewerage systems has been the structural health monitoring of concrete pipes to prevent potential leakages. In this context, a leakage detection system specifically designed for sewerage pipes and based on fiber optics has been reported and tested for five months [88]. Using FBGs as a relative humidity sensing element, the system demonstrated satisfactory performance for detecting cracks at the bottom of pipes. Fiber optic pH sensors have the potential to play an important role in sewerage monitoring, as pH is a clear indicator of biological and chemical corrosion. Measuring pH using fiber optics has been extensively studied, particularly in harsh environments. Thus, efficient and dedicated systems for monitoring the degradation of concrete sewerage structures, considered a main public concern, can be tackled using fiber optic sensors.

Other reported approaches of fiber optic sensing specifically designed for wastewater monitoring have used distributed temperature sensing [15, 89]. The basic idea is to detect the local change in the temperature of pipe walls caused by modifications in the flow conditions. These approaches have been tested in Austria and UK public sewer systems. Spatial resolutions in the order of meters and temperature resolutions of tenths of centigrade can be attained with fiber systems that may run along several kilometers. However, challenges in practical system applications can stem from temperature variations along the pipeline over extended distances and the positioning of the fiber cable within the actual pipeline.

5.4 Air monitoring

Tracking of air quality is another important parameter to evaluate in many fields, which in some cases requires constant monitoring of threshold levels of a specific substance. Activity from human settlements produces different pollutants that have consequences for the environment and living creatures. Particulate matter (PM) is undoubtedly a constant threat in big cities and highly industrial locations as a consequence of human activity. Environmental aftereffects of PM alter the incoming radiation on the Earth's surface, resulting in global warming [90]. Another repercussion of this pollutant is its effects on living creatures and humankind, causing breathing complications and disorders [90, 91]. Classification of these particles depends on their size and they are mainly grouped into coarse particles with diameters up to 10 μm (PM_{10}), fine particles with diameters up to 2.5 μm ($PM_{2.5}$), fine particles with diameters up to 1 μm (PM_1) and ultrafine particles below 1 μm [92, 93]. The most advanced techniques for PM detection rely on optical techniques based on scattering effects [94]. For this, changes in a laser source on airflow streams are evaluated and associated with the size of the pollutant [95].

Pollution assessment is thus done depending on the detection of the PM particles per air volume. The use of optical fibers in this instance is limited to confinement and direction of the laser source, resulting in very compact systems [96–98].

Excluding risks associated with particle size, detecting different substances is crucial for environmental problems. Although silica is ideal due to its low reactivity to different compounds, it is also a low porosity material, and hence, it is difficult to detect particles in the air through diffusion processes. Additionally, the low concentration and small size of particles in air make it more challenging to detect. Detection of substances or reagents in the air usually involves the interaction with gasses of different nature, which can be achieved in several ways depending on the type of contaminant.

Clearly, detection of contaminants or gasses with optical fibers suggests a direct interaction with the light guiding structure of the fiber [99–101]. Therefore, the most obvious approach takes advantage of microstructured (μOF) or photonic crystal (PCF) fibers, where the gas to be analyzed is injected within the microstructure of the fiber [101, 102]. For such a purpose, the inner structure of the fiber is highly important, as it can have a hollow or a solid core, and the former allows direct interaction with the gas and hence increased sensitivity [103, 104]. Some gasses or substances present absorption in the near-infrared; therefore, immediate analysis can be done while the gas is inserted in the fiber, as depicted in figure 5.7(a).

Commonly, direct detection of gasses can be done inside the μOF through absorption measurements in the NIR wavelengths, as these can be readily transmitted by the fibers [105, 106]. As a result, the fiber's transmission changes upon the

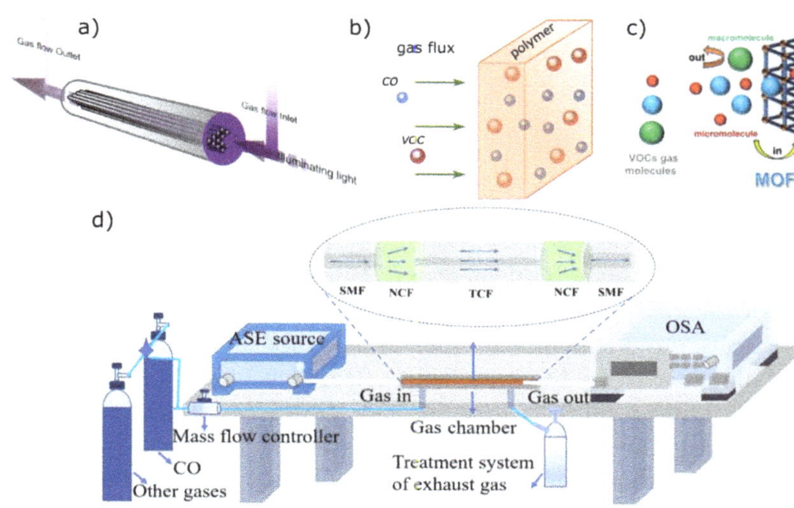

Figure 5.7. General mechanisms for gas detection in optical fibers. (a) Microstructured or photonic crystal fiber gas detection using direct gas filling (reproduced from [99] CC BY 4.0). Examples of materials added as a coating in optical fibers for gas detection and sensing are (b) a polymer layer presenting adsorption of a desired gas and (c) the use of metal–organic frameworks for gas capturing (reprinted from [113], copyright (2023), with permission from Elsevier). (d) An example of the basic experimental setup for gas detection with optical fibers (in this case, CO) (reprinted from [114], copyright (2022), with permission from Elsevier).

light's interaction with the gas, enabling direct monitoring through modification in the transmission spectra. Therefore, spectroscopy analysis inside the fiber can be readily performed [102, 106–108]. This method represents a straightforward and simple solution, although it involves some complexity because the gasses must be fed into the fiber. In addition, air quality monitoring and detecting concentrations of oxygen and humidity are readily achievable with this approach. For substances not showing specific absorption peaks within the operating spectral region, changes in the refractive index can be associated with the concentration of a known contaminant [109].

Solid structure optical fibers require the addition of other sensitive materials for detecting the contaminant of interest (see figure 5.7). Gasses of different natures can be monitored using sensing mechanisms such as fluorescence, interference, or evanescent wave interaction, as mentioned earlier in this chapter [99, 100, 110]. Such techniques are applicable to different types of gasses, which either react with the sensitive material or modify an optical feature of the fiber device, and the identification and quantification of the gas is specific to the sensitive material incorporated in the fibers. Conventional calibration and characterization of fiber-based gas sensors use chambers containing the sensor itself, and the gasses are injected in a controlled way (figure 5.7(d)). The sensor's response is acquired either in transmission or reflection configuration for different concentrations and flow conditions of the gasses.

In general, the most abundant emission gasses in the air are carbon monoxide (CO) and dioxide (CO_2) [111]. Other relevant greenhouse gasses are nitrous oxide (N_2O), methane (CH_4), ozone (O_3), and fluorinated gasses [112]. The advent of diversified and increasing industrial processes has further led to the production of volatile organic compounds (VOCs) such as ethanol, acetone, toluene, formaldehyde, and isopropyl alcohol, all of them a risk for health and the environment [113]. Gasses posing high risks for explosion, such as hydrogen (H), are also targeted in environmental detection systems.

5.4.1 Carbon oxides (CO and CO_2)

Carbon monoxide (CO) is a colorless and odorless gas that does not cause irritation when inhaled. The health risks arising from this gas are due to its binding to hemoglobin instead of oxygen and causing hypoxia [111]. CO usually comes from the inefficient combustion of hydrocarbons, and given its high toxicity, detection of this gas in the air is highly relevant. Concentrations above 25–50 ppm are considered dangerous and require high-sensitivity detection technologies [115].

Different materials or compounds reactive to CO can be incorporated on optical fiber sensors [100]. The most commonly adapted sensing mechanism uses catalytic effects generated on coatings incorporating metals over different sections of the fibers. Interaction of the gas with the coating can then be detected, for example, using interference effects. In [116], an α-Fe_2O_3/MgO layer was deposited directly on a section of an endlessly single-mode photonic crystal fiber coupled to a four-core fiber inside a Sagnac interferometer, whereas in [117], a polyaniline matrix containing Co_3O_4/CuO particles served as a coating on a thin-core fiber yielding a modal interferometer. MOFs are also very attractive for gas sensing due to their selectivity: structures such as Ni-MOF-74 and Ag/Co-MOF have been reported for CO sensing [118, 119]. These have been used as coatings on a tapered microfiber for evanescent wave interaction, as well as on a thin-core fiber incorporated in a Michelson interferometer. Some of these metal-based materials generate exothermic reactions, which can also be used for sensing. This approach was demonstrated using a coated capillary to create an FPI whose refractive index changes while interacting with CO [120]. Usually, these approaches are highly sensitive and show a good response in the presence of a few tens of ppm of CO. Other alternatives for CO sensors using materials without metal traces adsorb the gas and modify their refractive index. For example, cerium oxide and carbon nanotubes have been used in modal interferometers detecting up to 200 ppm in tapered fiber sections [114, 121].

Carbon dioxide (CO_2) is the main greenhouse gas resulting from burning fossil fuels and breathing. Constant monitoring of this gas is highly relevant for counteracting climate change [122]. Although its toxicity levels are much lower than those of CO gasses, ranging from hundreds to a few thousand ppm, it still represents a risk [123]. The most popular method to detect CO_2 emissions using optical fiber-based devices is by adding reactive fluorescent dyes on the tips of the fibers. The fluorescence intensity changes as a function of the CO_2 concentration, and this can be monitored with a photodiode or a spectrometer. pH-sensitive dyes such as HTPS, HEMA, acrylamide, methyl red, thymol blue, and phenol red, can be directly incorporated on the optical fibers, and an additional membrane permeable to CO_2 may be added to this arrangement [124–126]. Reactive dyes may also be immobilized in porous polymeric matrices, such as sol–gels or xerogels [127]. In pH-reactive dyes, wavelength selective attenuation can be tracked as a function of CO_2 concentration, or their response may be enhanced by adding quantum dots in the compound [128, 129]. Using this simple mechanism, multiple sensing-coated fibers can be bundled to form a multi-parameter sensor [129, 130]. Other alternatives to detect CO_2 gas include the use of photocrosslinkable poly ionic liquid with imidazolium functional groups to create Fabry–Pérot cavities or composites with plasmonic nanocrystals and hydrophobic zeolite embedded in a polymer matrix [131, 132].

Absorption measurements within different bands in the IR spectral region are also used for CO and CO_2 detection inside a μOF, particularly in the near IR wavelengths [133–135]. Since these gasses show stronger absorption bands towards the mid-IR wavelengths, μOF made of chalcogenide glass or other materials such as magnesium fluoride have been used for spectroscopic measurements, given their adequate transparency in this spectral region [136, 137].

5.4.2 Hazardous gasses and VOCs

VOCs are mainly composed of vapors from room-temperature volatile solvents. These are not only pollutants for the environment and hazardous for living species but also a potential explosion risk. Concentrations of a few to hundreds of ppm of these solvents can affect human health, causing dizziness, nausea, headaches, skin and eye irritation, respiratory complications, and, in the long term, even cancer [138]. The most common VOCs include alcohol vapors (methanol, ethanol, and isopropanol), acetone, benzene, formaldehyde, toluene, xylene, and acetylene, just to mention a few [100, 113, 138]. Fibers for VOCs detection usually require materials such as polymers, metallic composites, carbon-based materials, cholesteric liquid crystals, and coordinated compounds [113].

Polymeric materials are highly sensitive to physical and chemical factors, given their structure formed by chains of monomers. Thus, gasses and vapors can interact by permeating inside their structure and modifying either their volume or conductivity. Some of the most used polymers for VOC sensing are polydimethylsiloxane (PDMS), polyvinyl alcohol (PVA), polymethyl methacrylate (PMMA), polypyrrole (PPy), polyaniline (PANI), and polymeric liquid crystals (LCs) [113]. PPy and PANI are widely used because their conductivity is modified by redox reactions in the presence of different VOCs. PDMS is also recurrently used in optical fiber applications due to its optical characteristics and high permeability to nonpolar solvents that modify its refractive index [113, 138–140]. Polymers are a good option to coat different fiber sections, either for evanescent wave sensing or to create end caps forming resonant cavities [99, 113, 138]. For example, a side polished fiber incorporating polymer coatings with organic dyes has shown to be useful as a colorimetric sensor for pollutants [141].

Detection of alcohol vapors has been achieved by incorporating different materials as the cladding for the fibers. Polymers can be used as the cladding layer or as coatings for tapered sections of fibers. A vast amount of reports use metal oxides for detection, such as V_2O_5, WO_3, Bi_2O_3, ZnO, (TiO_2)-doped rhodamine 6G, and $MnCo_2O_4$ nanocrystals. In general, the interaction with the vapors produces changes in the material's refractive index, which modifies the absorption coefficients [113]. These metal oxides are usually incorporated into fiber sections, and the vapors interact with the evanescent wave. For instance, adsorption of the VOC on layers of Au and Ag metal–organic complexes has been shown to increase the attenuation of the light transmitted by the coated fiber [142, 143]. Similarly, optical fibers made of silver halide (AgBrCl) operating in the mid-IR have used a coating of nanoporous silicon particles to detect VOCs [144]. Other materials used to detect these compounds include nano-amorphous SnO_2 layers [113].

Benzene is a product directly produced by the petroleum industry, and also from certain polymer and resin synthesis. It is extremely hazardous for the environment and can affect the nervous and immune systems of humans. Despite its toxicity, only a few reports of benzene sensors using optical fibers can be found in the literature. The main approach is based on the use of PCFs to allocate the chemicals in the hollow structure of these fibers and track the spectral changes in transmission [138]. A PCF coated with carbon nanotubes and spliced to SMFs forming an MZ device has also shown good sensitivity to this VOC [145]. Calixarene has also been used as a coating on an LPG, achieving high sensitivity and specificity to benzene and aromatic compounds [146].

Ammonia (NH_3) is a highly toxic gas widely produced in many industrial processes and used as fertilizer. A highly selective NH_3 sensor was reported using silica gel as the coating material for a microfiber coupler [147]. Evanescent interaction with the coating changed the coupling coefficient due to the chemo absorption of the silica gel. Metal oxides, such as (Sm_2O_3), have also been reported to be sensitive to this VOC and alcohol vapors [148].

5.4.3 Soil and mines monitoring

Monitoring flammable or explosive gasses such as H_2, NO_2, H_2S, NH_3, CH_4, CO, and CO_2 in mines, tunnels, excavations, and landfills is of great interest as the risk of accumulation in these locations is critical [149]. Many of these are odorless gasses and highly toxic in small concentrations, and their accumulation can be easily unnoticed, which increases the risk of explosions. The use of optical fiber sensing technology for monitoring this type of gasses minimizes the risks of explosions mainly because fiber sensors do not require electrical signals [149]. The most common risks for soil contamination are associated with natural reservoirs of CO_2. A multimodal interference sensor using a hollow core PCF spliced to SMF has been shown to be useful for monitoring CO_2 concentration [150]. Other sources of contamination come from VOCs due to industrial sources, and coatings of polymeric materials with dyes have been used to detect volatile hydrocarbons [151].

5.5 Dangerous and extreme environments

In the context of environmental applications, monitoring physical parameters and/or detecting chemical species in harsh conditions is mostly relevant for safety reasons. As an example, detecting liquid spills and gas leakages in nuclear power plants, or oil and gas wells is paramount to activating suitable safety procedures. In some cases, sensors have to withstand conditions such as exposure to radiation and or corrosive media under extreme temperature conditions [152]. In addition to safety, environmental monitoring can also serve to efficiently handle products obtained from resources such as mines, natural gas, oil, and cultivated soil [153]. The harsh conditions commonly found in these applications impose stringent requirements for the design, fabrication, and material selection of the sensing devices. Silica-based fiber sensors might be treated to enhance their thermal stability, or other materials, such as sapphire, can also be used for device fabrication. The use of metallic, polymers, oxides, or carbon coatings can also enhance the performance of the fibers [152]. Optical fiber sensors allow for monitoring parameters at reduced spatial scales (i.e., point sensing, typically a few cm) or in quasi-distributed arrays (multiple point sensing) covering extended lengths. However, the most attractive feature of fiber sensor technology is the capability to perform distributed sensing, allowing for monitoring parameters over tens of kilometers.

As described in a previous section, scattering-based sensors are generally used for distributed sensing, yielding temperature and/or strain profiles along the optical fiber. The sensing element is the fiber itself, and SMF or MMF can be used. Cables containing several fibers have been deployed to obtain temperature and strain maps

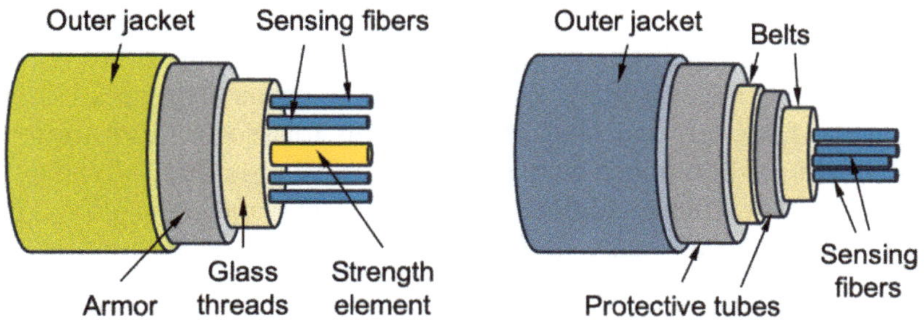

Figure 5.8. Examples of fiber optic cable layouts for sensing: near-surface sensing (left) and down-hole sensing (right). A cable typically contains several fibers for sensing; for high-temperature applications, the sensing fibers are generally coated with high-temperature-resistant polymers such as polyimide.

in oil and gas reservoirs, industrial settings, and offshore platforms. The cable design will vary depending on the deployment location (near-surface, down-hole, high temperature, see figure 5.8). Although the spatial resolution, sensitivity, and accuracy of these sensors depend on the interrogation scheme, fiber optic distributed sensing can readily yield resolutions of tens of centimeters over lengths of several tens of km [13, 154]. Improvements in the performance of these distributed sensing techniques are based on signal processing, light source modulation schemes, and multi-pumping experimental schemes [154]. As an example, novel interrogation strategies along with signal processing techniques, have enabled distributed acoustic sensing (DAS) for remote monitoring of earthquakes and landslides (see the previous sections of this chapter). Distributed sensing can further be used in mining activities to assess overburden deformation, which may lead to fractured underground zones impacting the surface structure stability and altering the ecological balance of the exploited zone [155]. Additionally, volatile organic compounds may also be monitored with optical fiber point sensors [156], thus offering a suitable sensing platform for the extreme conditions commonly found in mines, and gas and oil wells [157, 158].

5.5.1 High temperature

Although the melting point of pure silica is around 1700 °C, noticeable changes in the material of standard optical fibers (doped silica) start at temperatures close to 300 °C [152]. This arises through diffusion processes of the dopants or changes in the refractive index with temperature (i.e., the thermo-optic coefficient of the material) [159, 160]. Before sensor fabrication, the fibers can be annealed to increase their temperature stability; alternatively, temperature-resistant coatings (e.g., polyimide, metals, or carbon) can be applied to increase the temperature resistance and prevent diffusion of liquids and gasses into the glass material [152, 159, 160]. While polyimide coatings, for instance, can provide extended lifetimes of the fibers at temperatures beyond 300 °C, metallic coatings using nickel, copper, aluminum, or gold have been shown to yield thermal stability up to 600 °C [152]. An increased hermetic protection is achieved with thin coatings of carbon (20–50 nm) deposited onto the fiber cladding during the drawing process, preventing diffusion of water and hydrogen, thereby reducing the losses and degradation induced by these molecules [161]. For temperatures beyond 600 °C, sapphire fibers are the preferred choice because the melting point of this material extends up to 2050 °C [162]. Sapphire fibers, however, impose other constraints for sensor fabrication: they are not as ubiquitous as silica fibers, increasing the sensors' cost. Typically, they are used as sensing heads for extremely high-temperature sensors, and conventional fibers can deliver the light for interrogation. Sapphire fibers can also be used to construct devices with multilayered ceramic materials yielding interferometric sensors (e.g., Fabry–Pérot) [163].

The detrimental effects of elevated temperatures on the structure of the glass evidently have an impact on the different fiber devices used as sensors. Fiber gratings (both FBGs and LPFGs) will experience thermal drift, and more importantly, they can be permanently 'erased' when exposed to temperatures close to 300 °C for extended periods [152]. Once again, the use of special coatings provides a means to extend the temperature range, albeit potentially affecting their sensitivity. Special coatings can also be used in scattering-based fiber sensors to obtain distributed measurements for temperatures beyond the 300 °C limit [164]. The temperature performance of some fiber devices, such as FBGs, can also vary depending on their fabrication method. Type II FBGs, for instance, fabricated with pulsed lasers under various exposure conditions can perform well at temperatures of up to 600 °C [152]. Short-pulse laser processing of silica fibers can also provide a means to fabricate sensing devices such as Fabry–Pérot cavities with all-glass structures capable of withstanding up to 600 °C [165]. Higher temperature ranges have also been achieved with intrinsic F–P sensors fabricated in the core of the optical fiber using a femtosecond pulsed laser [166]. The changes in the fiber's refractive index induced by the femtosecond laser are stable for temperature ranges extending beyond 1000 ° C, yielding sensors that can withstand several temperature cycles. Furthermore, since they are wavelength-encoded, the devices can be used in multiplexed arrays for multi-point sensing applications [166].

5.5.2 Cryogenic temperatures

Monitoring temperatures at cryogenic levels is relevant in applications such as the storage of liquefied natural gas or the energy sector, particularly in nuclear reactors [152]. In general, the materials used to fabricate the optical fiber will impose practical limits on the performance of the sensors. As discussed in the previous section, the thermal properties of the doped silica and the coating material will both influence the temperature response. While thermal expansion plays a role for some devices (e.g., fiber gratings and F–P sensors), the dominant effects are related to the change in refractive index, described by the thermo-optical coefficients of the glass [152]. The thermo-optic coefficient accounts for the change in refractive index with temperature, and changes due to stresses generated through contraction or expansion of the glass are covered by the thermo-elastic coefficient. These coefficients are linear over a reduced range within room temperature conditions, but become nonlinear as the temperature decreases and almost vanishes for extremely low temperatures ($T \leqslant 50$ K) [167]. A common practice to overcome this limitation is using coatings with suitable mechanical and thermal properties to increase the operational range. Acrylate, PMMA, ceramic, and metallic coatings are the most commonly used materials, as they have larger thermal expansion coefficients and Young's moduli larger than silica glass [152, 167, 168]. Aside from the intrinsic properties of the material, the coating thickness will also play a role in the sensor's performance.

The strategy of using coatings to increase the operational range of fiber sensors has been widely used for point sensors and scattering-based sensing schemes. Devices such as FBGs sensors incorporating metals, ceramics, or combined layers of different materials as coatings have been shown to perform well at temperatures as low as 4 K [24, 25]. Hence, multiplexed sensor arrays for multi-point sensing schemes can be realized for extremely low-temperature applications. Similarly, distributed sensing can be performed using proper coatings along the sensing fiber using the traditional back-reflected schemes [167, 168].

5.5.3 High radiation environments

High radiation levels (gamma rays, x-rays, neutrons) cause several effects in silica glass and thus affect the performance of optical fiber sensors. Radiation-induced defects in the glass structure (core and cladding) can lead to attenuation, thereby reducing light transmission. In addition to this radiation-induced darkening, other effects owing to radiation exposure of the glass include radiation-induced emission (e.g., radioluminescence) and radiation-induced refractive index change (also termed compaction) [152, 169]. Although, in some cases, these effects might help sense applications (emission can be used for dosimetry), efforts are made to mitigate them to extend the use of fiber sensors under high radiation conditions. Some applications may also require monitoring elevated temperatures; hence, techniques such as those described previously (i.e. special coatings and other materials such as sapphire [170]) might also be applicable.

Metallic coatings and high-temperature resistance polymers (e.g., polyimide) have been reportedly useful in optical fiber sensors exposed to high radiation environments under elevated temperatures [152]. However, the use of special fibers such as all-silica fibers or silica with different dopants in the core or the cladding (or both) is the preferred choice [152, 169, 171]. Point sensors such as FBGs or FPI, as well as scattering-based sensors, have been successfully implemented with these fibers tailored to withstand high radiation levels. Doses as high as 10 MGy have been reportedly supported by FBGs, allowing one to monitor parameters such as temperature and stress in multi-point configurations in nuclear reactors.

5.5.4 Oil and gas industry

The use of fiber optic sensor technology for the oil and gas industry has been explored for several decades, and systems based on distributed and point sensors have already been deployed and tested. Since the first installation of a fiber pressure sensor in the early 1990s in an oil well [172], fiber sensor technology has increasingly found applications in the three major sectors of the oil and gas industry (i.e., upstream, midstream, and downstream) [173]. The range of activities in these sectors spans from geological surveys and seismic acquisition to flow and leak detection of liquids and gasses. Hence, optical fiber sensors have proven to be a versatile technology capable of performing well in all the harsh conditions found during oil and gas production. In general, the information provided by the sensors is used to optimize the extraction and production of oil, gas, and other hydrocarbon products. Evidently, optimization of these processes is relevant as it has a direct impact on the environment. Furthermore, monitoring pressure, temperature, and flow across the different stages of oil and gas production also provides a means to detect leaks of fluids that can directly impact the environment.

Point sensing schemes have been successfully demonstrated for monitoring pressure, temperature, and seismic activities in oil wells. These are typically based on FBGs and Fabry–Pérot resonators in intrinsic or extrinsic configurations, allowing multiplexed schemes using several devices along a single fiber (i.e., multipoint sensing) [174, 175]. However, the most attractive feature offered by fiber optic technology is its distributed sensing capabilities, as it allows for obtaining temperature and seismic 'maps' directly from the wells into which the fiber cable is deployed [174]. As discussed in the previous sections, distributed sensing is based on scattering effects within the fiber, allowing one to measure parameters over tens of kilometers in real-time with spatial resolutions under 0.5 m [173, 175]. Two techniques are the most commonly used for these applications: distributed temperature sensing (DTS) and DAS. The former is based on Raman scattering, while the latter relies on Rayleigh scattering. DTS has become a well-established technique for mapping temperatures across the three sectors of the oil and gas industry, and DAS has increasingly found applications in different stages of oil and gas production. For example, hydraulic fracturing processes have been monitored using DAS technology [173]. Strategies have also been devised to simultaneously analyze more than one scattering effect along the fiber, yielding novel techniques such as distributed temperature strain sensing (DTSS). In this case, in addition to obtaining information about the temperature along the fiber through the Raman effect, strain measurements can be performed through Brillouin backscattering analysis [164, 175].

References

[1] Muñoz-Liesa J, Toboso-Chavero S, Mendoza Beltran A, Cuerva E, Gallo E, Gassó-Domingo S and Josa A |2021 Building-integrated agriculture: are we shifting environmental impacts? An environmental assessment and structural improvement of urban greenhouses *Resour. Conserv. Recycl.* **169** 105526

[2] Molitor D, Mullins J T and White C 2023 Air pollution and suicide in rural and urban America: evidence from wildfire smoke *PNAS* **120** e2221621120

[3] Rao C and Yan B 2020 Study on the interactive influence between economic growth and environmental pollution *Environ. Sci. Pollut. Res.* **27** 39442–65

[4] Başar S and Tosun B 2021 Environmental Pollution Index and economic growth: evidence from OECD countries *Environ. Sci. Pollut. Res.* **28** 36870–9

[5] Pendão C and Silva I 2022 Optical fiber sensors and sensing networks: overview of the main principles and applications *Sensors* **22** 7554
[6] Korposh S, James S W, Lee S W and Tatam R P 2019 Tapered optical fibre sensors: current trends and future perspectives *Sensors (Basel)* **19** 2294
[7] Lee B H, Kim Y H, Park K S, Eom J B, Kim M J, Rho B S and Choi H Y 2012 Interferometric fiber optic sensors *Sensors* **12** 2467–86
[8] Hill K O and Meltz G 1997 Fiber Bragg grating technology fundamentals and overview *J. Light. Technol.* **15** 1263–76
[9] Bhaskar C V N, Pal S and Pattnaik P K 2021 Recent advancements in fiber Bragg gratings based temperature and strain measurement *Results Opt.* **5** 100130
[10] Li C, Tang J, Cheng C, Cai L and Yang M 2021 FBG arrays for quasi-distributed sensing: a review *Photonic Sens.* **11** 91–108
[11] Cibula E and Donlagic D 2007 In-line short cavity fabry-perot strain sensor for quasi distributed measurement utilizing standard OTDR *Opt Express* **15** 8719–30
[12] Lu P, Lalam N, Badar M, Liu B, Chorpening B T, Buric M P and Ohodnicki P R 2019 Distributed optical fiber sensing: review and perspective *Appl. Phys. Rev.* **6** 041302
[13] Palmieri L, Schenato L, Santagiustina M and Galtarossa A 2022 Rayleigh-based distributed optical fiber sensing *Sensors* **22** 6811
[14] Mondanos M, Parker T, Milne C H, Yeo J, Coleman T and Farhadiroushan M 2015 *Distributed temperature and distributed acoustic sensing for remote and harsh environments***vol 9491** ed D G Senesky and S Dekate (SPIE) 94910F
[15] Apperl B, Bernhardt M and Schulz K 2015 Distributed temperature sensing (DTS) als Messverfahren in Landoberflächenhydrologie und Siedlungswasserwirtschaft *Österr. Wasser- Abfallwirtsch.* **67** 447–56
[16] Dunkerley D 2022 Acoustic methods in geophysics *Preview* **2022** 42–7
[17] Zhan Z 2019 Distributed acoustic sensing turns fiber-optic cables into sensitive seismic antennas *Seismol. Res. Lett.* **91** 1–15
[18] Fernández-Ruiz M R, Soto M A, Williams E F, Martin-Lopez S, Zhan Z, Gonzalez-Herraez M and Martins H F 2020 Distributed acoustic sensing for seismic activity monitoring *APL Photonics* **5** 030901
[19] Lindsey N J, Martin E R, Dreger D S, Freifeld B, Cole S, James S R, Biondi B L and Ajo-Franklin J B 2017 Fiber-optic network observations of earthquake wavefields *Geophys. Res. Lett.* **44** 11792–9
[20] Wang X, Williams E F, Karrenbach M, Herráez M G, Martins H F and Zhan Z 2020 Rose parade seismology: signatures of floats and bands on optical fiber *Seismol. Res. Lett.* **91** 2395–8
[21] Yu J, Xu P, Yu Z, Wen K, Yang J, Wang Y and Qin Y 2023 Principles and applications of seismic monitoring based on submarine optical cable *Sensors* **23** 5600
[22] Lior I, Rivet D, Ampuero J P, Sladen A, Barrientos S, Sánchez-Olavarría R, Villarroel Opazo G A and Bustamante Prado J A 2023 Magnitude estimation and ground motion prediction to harness fiber optic distributed acoustic sensing for earthquake early warning *Sci. Rep.* **13** 424
[23] Bogris A *et al* 2022 Sensitive seismic sensors based on microwave frequency fiber interferometry in commercially deployed cables *Sci. Rep.* **12** 14000
[24] Noe S, Husmann D, Müller N, Morel J and Fichtner A 2023 Long-range fiber-optic earthquake sensing by active phase noise cancellation *Sci. Rep.* **13** 13983

[25] Zhan Z, Cantono M, Kamalov V, Mecozzi A, Müller R, Yin S and Castellanos J C 2021 Optical polarization-based seismic and water wave sensing on transoceanic cables *Science (1979)* **371** 6532
[26] Hernandez P D, Ramirez J A and Soto M A 2022 Deep-learning-based earthquake detection for fiber-optic distributed acoustic sensing *J. Lightwave Technol.* **40** 2639–50
[27] Saad O M *et al* 2023 Earthquake forecasting using big data and artificial intelligence: a 30-week real-time case study in China *Bull. Seismol. Soc. Am.* **113** 2461–78
[28] Currenti G, Jousset P, Napoli R, Krawczyk C and Weber M 2021 On the comparison of strain measurements from fibre optics with a dense seismometer array at Etna volcano (Italy) *Solid Earth* **12** 993–1003
[29] Jousset P, Currenti G, Schwarz B, Chalari A, Tilmann F, Reinsch T, Zuccarello L, Privitera E and Krawczyk C M 2022 Fibre optic distributed acoustic sensing of volcanic events *Nat. Commun.* **13** 1753
[30] Nishimura T, Emoto K, Nakahara H, Miura S, Yamamoto M, Sugimura S, Ishikawa A and Kimura T 2021 Source location of volcanic earthquakes and subsurface characterization using fiber-optic cable and distributed acoustic sensing system *Sci. Rep.* **11** 6319
[31] Biagioli F, Métaxian J-P, Stutzmann E, Ripepe M, Bernard P, Trabattoni A, Longo R and Bouin M-P 2024 Array analysis of seismo-volcanic activity with distributed acoustic sensing *Geophys. J. Int.* **236** 607–20
[32] Currenti G *et al* 2023 Distributed dynamic strain sensing of very long period and long period events on telecom fiber-optic cables at Vulcano, Italy *Sci. Rep.* **13** 4641
[33] Klaasen S, Thrastarson S, Çubuk-Sabuncu Y, Jónsdóttir K, Gebraad L, Paitz P and Fichtner A 2023 Subglacial volcano monitoring with fiber-optic sensing: Grímsvötn, Iceland *Volcanica* **6** 301–11
[34] Fichtner A, Klaasen S, Thrastarson S, Çubuk-Sabuncu Y, Paitz P and Jónsdóttir K 2022 Fiber-optic observation of volcanic tremor through floating ice sheet resonance *Seismic Rec.* **2** 148–55
[35] Auflič M J *et al* 2023 Landslide monitoring techniques in the Geological Surveys of Europe *Landslides* **20** 951–65
[36] Zhu H H, Shi B and Zhang C C 2017 FBG-based monitoring of geohazards: current status and trends *Sensors (Switzerland)* **17** 452
[37] Pei H, Cui P, Yin J, Zhu H, Chen X, Pei L and Xu D 2011 Monitoring and warning of landslides and debris flows using an optical fiber sensor technology *J. Mt. Sci.* **8** 728–38
[38] Zheng Y, Zhu Z W, Deng Q X and Xiao F 2019 Theoretical and experimental study on the fiber Bragg grating-based inclinometer for slope displacement monitoring *Opt. Fiber Technol.* **49** 28–36
[39] Zhang L, Shi B, Zeni L, Minardo A, Zhu H and Jia L 2019 An fiber Bragg grating-based monitoring system for slope deformation studies in geotechnical centrifuges *Sensors (Switzerland)* **19** 1591
[40] Peng H, Chen B, Dong P, Chen S, Liao Y and Guo Q 2020 Application of FBG sensing technology to internal deformation monitoring of landslide *Adv. Civil Eng.* **2020** 1238945
[41] Liu Y, Dai Z, Zhang X, Peng Z, Li J, Ou Z and Liu Y 2010 Optical fiber sensors for landslide monitoring *Proc. Semiconductor Lasers and Applications IV* vol 7844 78440D
[42] Schenato L, Palmieri L, Camporese M, Bersan S, Cola S, Pasuto A, Galtarossa A, Salandin P and Simonini P 2017 Distributed optical fibre sensing for early detection of shallow landslides triggering *Sci. Rep.* **7** 14686

[43] Iten M, Puzrin A M and Schmid A 2008 Landslide monitoring using a road-embedded optical fiber sensor *Proc. Smart Sensor Phenomena, Technology, Networks, and Systems* vol 6933 693315

[44] Minardo A, Zeni L, Coscetta A, Catalano E, Zeni G, Damiano E, De Cristofaro M and Olivares L 2021 Distributed optical fiber sensor applications in geotechnical monitoring *Sensors* **21** 7514

[45] Gutiérrez F, Sevil J, Sevillano P, Preciado-Garbayo J, Martínez J J, Martín-López S and González-Herráez M 2023 The application of distributed optical fiber sensors (BOTDA) to sinkhole monitoring. Review and the case of a damaging sinkhole in the Ebro Valley evaporite karst (NE Spain) *Eng. Geol.* **325** 107289

[46] UNESCO 2009 Water in a Changing World *The United Nations World Water Development Report 3* (UNESCO) https://unesdoc.unesco.org/ark:/48223/pf0000181993

[47] Uddin M G, Nash S and Olbert A I 2021 A review of water quality index models and their use for assessing surface water quality *Ecol. Indic.* **122** 107218

[48] 2013 *Smart Sensors for Real-Time Water Quality Monitoring* Smart Sensors, Measurement and Instrumentation C Mukhopadhyay and A Mason (Springer)

[49] Jderu A, Dorobantu D, Ziegler D and Enachescu M 2022 Swelling-based chemical sensing with unmodified optical fibers *Photonic Sensors* **12** 99–104

[50] Lyons W B, Ewald H, Flanagan C and Lewis E 2003 A multi-point optical fibre sensor for condition monitoring in process water systems based on pattern recognition *Measurement* **34** 301–12

[51] Lewis E, Sheridan C, O'Farrell M, King D, Flanagan C, Lyons W B and Fitzpatrick C 2007 Principal component analysis and artificial neural network based approach to analysing optical fibre sensors signals *Sensors Actuators* A **136** 28–38

[52] Irigoyen M, Sánchez-Martin J A, Bernabeu E and Zamora A 2017 Tapered optical fiber sensor for chemical pollutants detection in seawater *Meas. Sci. Technol.* **28** 045802

[53] Raichlin Y, Marx S, Gerber L and Katzir A 2004 Infrared fiber optic evanescent wave spectroscopy and its applications for the detection of toxic materials in water, *in situ* and in real time *Proc. Optically Based Biological and Chemical Sensing for Defence* vol 5617J C Carrano and A Zukauskas (SPIE) p 145

[54] McCue R P, Walsh J E, Walsh F and Regan F 2006 Modular fibre optic sensor for the detection of hydrocarbons in water *Sens. Actuators B Chem.* **114** 438–44

[55] Steiner H *et al* 2003 *In situ* sensing of volatile organic compounds in groundwater: first field tests of a mid-infrared fiber-optic sensing system *Appl. Spectrosc.* **57** 607–13

[56] Wang M, Dai S, Gan N and Wang Y 2022 *In situ* growth of silver nanoparticles on polydopamine-coated chalcogenide glass tapered fiber for the highly sensitive detection of volatile organic compounds in water *J. Non Cryst. Solids* **581** 121420

[57] Cennamo N, Arcadio F, Perri C, Zeni L, Sequeira F, Bilro L, Nogueira R, D'Agostino G, Porto G and Biasiolo A 2019 Water monitoring in smart cities exploiting plastic optical fibers and molecularly imprinted polymers. The case of PFBS detection *2019 IEEE Int. Symp. on Measurements & Networking (M&N)* (IEEE) pp 1–6

[58] Fallah H, Asadishad T, Parsanasab G M, Harun S W, Mohammed W S and Yasin M 2021 Optical fiber biosensor toward *e-coli* bacterial detection on the pollutant water *Eng. J.* **25** 1–8

[59] Yildirim N, Long F, Gao C, He M, Shi H-C and Gu A Z 2012 Aptamer-based optical biosensor for rapid and sensitive detection of 17β-estradiol in water samples *Environ. Sci. Technol.* **46** 3288–94

[60] Nguyen L V, Vasiliev M and Alameh K 2011 Three-wave fiber Fabry–Pérot interferometer for simultaneous measurement of temperature and water salinity of seawater *IEEE Photonic. Technol. Lett.* **23** 450–2

[61] Pereira D A 2004 Fiber Bragg grating sensing system for simultaneous measurement of salinity and temperature *Opt. Eng.* **43** 299

[62] Yang H, Wang S, Wang X, Wang J and Liao Y 2014 Temperature sensing in seawater based on microfiber knot resonator *Sensors* **14** 18515–25

[63] Qian J, Lv R, Wang S, Zhao Y and Zhao Q 2018 High-sensitivity temperature sensor based on single-mode fiber for temperature-measurement application in the ocean *Opt. Eng.* **57** 1

[64] Qi X, Wang S, Jiang J, Liu K, Wang X, Yang Y and Liu T 2019 Fiber optic fabry-perot pressure sensor with embedded MEMS micro-cavity for ultra-high pressure detection *J. Lightwave Technol.* **37** 2719–25

[65] Bae H, Zhang X M, Liu H and Yu M 2010 Miniature surface-mountable Fabry–Perot pressure sensor constructed with a 45° angled fiber *Opt. Lett.* **35** 1701

[66] Wang K 2009 Seawater pH sensor based on the long period grating in a single-mode–multimode–single-mode structure *Opt. Eng.* **48** 034401

[67] Cheng X, Bonefacino J, Guan B O and Tam H Y 2018 All-polymer fiber-optic pH sensor *Opt. Express* **26** 14610

[68] Goicoechea J, Zamarreño C R, Matías I R and Arregui F J 2008 Optical fiber pH sensors based on layer-by-layer electrostatic self-assembled neutral red *Sens. Actuators* B **132** 305–11

[69] Lei M, Zhang Y-N, Han B, Zhao Q, Zhang A and Fu D 2018 In-line Mach–Zehnder interferometer and FBG with smart hydrogel for simultaneous pH and temperature detection *IEEE Sens.* J. **18** 7499–504

[70] Zhao Y, Lei M, Liu S-X and Zhao Q 2018 Smart hydrogel-based optical fiber SPR sensor for pH measurements *Sens. Actuators* B **261** 226–32

[71] Rustandi D, Prakosa J A, Purwowibowo P, Wijornako S, Maftukhah T, Kurniawan E, Sirenden B H and Mahmudi M 2023 Measurement uncertainty evaluation of salinity sensing through water electrical conductivity method with gravimetric validation *AIP Conf. Proc.* **2906** 040004

[72] Wang S, Liu T, Wang X, Liao Y, Wang J and Wen J 2019 Hybrid structure Mach-Zehnder interferometer based on silica and fluorinated polyimide microfibers for temperature or salinity sensing in seawater *Measurement* **135** 527–36

[73] Yang F, Hlushko R, Wu D, Sukhishvili S A, Du H and Tian F 2019 Ocean salinity sensing using long-period fiber gratings functionalized with layer-by-layer hydrogels *ACS Omega* **4** 2134–41

[74] Liao Y, Wang J, Yang H, Wang X and Wang S 2015 Salinity sensing based on microfiber knot resonator *Sens. Actuators* A **233** 22–5

[75] Aslam Mollah M, Yousufali M, Rifat Bin Asif Faysal M, Rabiul Hasan M, Hossain M B and Amiri I S Highly sensitive photonic crystal fiber salinity sensor based on Sagnac interferometer *Results Phys.* **16** 103022

[76] World Health Organization 2011 *Guidelines for Drinking-water Quality* (Geneva: World Health Organization)

[77] Kapoor D and Singh M P 2021 Heavy metal contamination in water and its possible sources *Heavy Metals in the Environment* (Amsterdam: Elsevier) pp 179–89

[78] Bhavsar K, Prabhu R and Pollard P 2013 Development of dithizone based fibre optic evanescent wave sensor for heavy metal ion detection in aqueous environments *J. Phys. Conf. Ser.* **450** 012011
[79] Bhavsar K, Hurston E, Prabhu R and Joseph G P 2017 Fibre optic sensor to detect heavy metal pollutants in water environments *OCEANS 2017—Aberdeen* (Piscataway, NJ: IEEE) pp 1–4
[80] Lin Y, Dong X, Yang J, Maa H, Zu P, So P L and Chan C C 2017 Detection of Ni^{2+} with optical fiber Mach-Zehnder interferometer coated with chitosan/MWCNT/PAA *2017 16th International Conference on Optical Communications and Networks (ICOCN)* (Piscataway, NJ: IEEE) pp 1–3
[81] Ji W B, Yap S H K, Panwar N, Zhang L L, Lin B, Yong K T, Tjin S C, Ng W J and Majid M B A 2016 Detection of low-concentration heavy metal ions using optical microfiber sensor *Sens. Actuators* B **237** 142–9
[82] Kishore P V N, Madhuvarasu S S and Moru S 2018 Stimulus responsive hydrogel-coated etched fiber Bragg grating for carcinogenic chromium (VI) sensing *Opt. Eng.* **57** 1
[83] Liu C, Sun Z, Zhang L, Lv J, Yu X F, Zhang L and Chen X 2018 Black phosphorus integrated tilted fiber grating for ultrasensitive heavy metal sensing *Sens. Actuators* B **257** 1093–8
[84] Tabassum R and Gupta B D 2015 Fiber optic manganese ions sensor using SPR and nanocomposite of ZnO–polypyrrole *Sens. Actuators* B **220** 903–9
[85] Shrivastav A M and Gupta B D 2018 Ion-imprinted nanoparticles for the concurrent estimation of Pb(II) and Cu(II) ions over a two channel surface plasmon resonance-based fiber optic platform *J. Biomed. Opt.* **23** 1–8
[86] Zhou M, Guo J and Yang C 2018 Ratiometric fluorescence sensor for Fe^{3+} ions detection based on quantum dot-doped hydrogel optical fiber *Sens. Actuators* B **264** 52–8
[87] Long F, Gao C, Shi H C, He M, Zhu A N, Klibanov A M and Gu A Z 2011 Reusable evanescent wave DNA biosensor for rapid, highly sensitive, and selective detection of mercury ions *Biosens. Bioelectron.* **26** 4018–23
[88] Alwis L S M, Bustamante H, Bremer K, Roth B, Sun T and Grattan K T V 2016 A pilot study: evaluation of sensor system design for optical fibre humidity sensors subjected to aggressive air sewer environment *2016 IEEE Sensors* (Piscataway, NJ: IEEE) pp 1–3
[89] Kechavarzi C, Keenan P, Xu X and Rui Y 2020 Monitoring the hydraulic performance of sewers using fibre optic distributed temperature sensing *Water (Basel)* **12** 2451
[90] Mukherjee A and Agrawal M 2017 World air particulate matter: sources, distribution and health effects *Environ. Chem. Lett.* **15** 283–309
[91] Anderson J O, Thundiyil J G and Stolbach A 2012 Clearing the air: a review of the effects of particulate matter air pollution on human health *J. Med. Toxicol.* **8** 166–75
[92] Amaral S, de Carvalho J, Costa M and Pinheiro C 2015 An overview of particulate matter measurement instruments *Atmosphere (Basel)* **6** 1327–45
[93] Hopke P K, Dai Q, Li L and Feng Y 2020 Global review of recent source apportionments for airborne particulate matter *Sci. Total Environ.* **740** 140091
[94] Su X, Sutarlie L and Loh X J 2020 Sensors and analytical technologies for air quality: particulate matters and bioaerosols *Chem. Asian J.* **15** 4241–55
[95] Molaie S and Lino P 2021 Review of the newly developed, mobile optical sensors for real-time measurement of the atmospheric particulate matter concentration *Micromachines (Basel)* **12** 416

[96] Wang Y and Muth J 2017 An optical-fiber-based airborne particle sensor *Sensors* **17** 2110
[97] Jobert G, Barritault P, Fournier M, Boutami S, Jobert D, Marchant A, Michelot J, Monsinjon P, Lienhard P and Nicoletti S 2020 Miniature particulate matter counter and analyzer based on lens-free imaging of light scattering signatures with a holed image sensor *Sens. Actuators Rep.* **2** 100010
[98] Sohn I-B, Choi H-K, Jung Y-J, Lee C-J, Oh M-K and Ahsan M S 2023 Measurement of fine/ultrafine dust using lenticular fiber-based particulate measurement devices *IEEE Sens. J.* **23** 8400–9
[99] Allsop T and Neal R 2021 A review: application and implementation of optic fibre sensors for gas detection *Sensors* **21** 6755
[100] Lyu D, Huang Q, Wu X, Nie Y and Yang M 2023 Optical fiber sensors for water and air quality monitoring: a review *Opt. Eng.* **63** 031004
[101] Li J, Yan H, Dang H and Meng F 2021 Structure design and application of hollow core microstructured optical fiber gas sensor: a review *Opt. Laser Technol.* **135** 106658
[102] Paul B K, Ahmed K, Dhasarathan V, Al-Zahrani F A, Aktar M N, Uddin M S and Aly A H 2020 Investigation of gas sensor based on differential optical absorption spectroscopy using photonic crystal fiber *Alex. Eng. J.* **59** 5045–52
[103] Cordeiro C M B, Franco M A R, Chesini G, Barretto E C S, Lwin R, Brito Cruz C H and Large M C J 2006 Microstructured-core optical fibre for evanescent sensing applications *Opt. Express* **14** 13056
[104] Nizar S M, Britto E C, Michael M and Sagadevan K 2023 Photonic crystal fiber sensor structure with vertical and horizontal cladding for the detection of hazardous gases *Opt. Quantum Electron.* **55** 1186
[105] Morshed M, Imran Hassan M, Roy T K, Uddin M S and Abdur Razzak S M 2015 Microstructure core photonic crystal fiber for gas sensing applications *Appl. Opt.* **54** 8637
[106] Mishra G P, Kumar D, Chaudhary V S and Kumar S 2022 Design and sensitivity improvement of microstructured-core photonic crystal fiber based sensor for methane and hydrogen fluoride detection *IEEE Sens. J.* **22** 1265–72
[107] Zhao P, Zhao Y, Bao H, Ho H L, Jin W, Fan S, Gao S, Wang Y and Wang P 2020 Mode-phase-difference photothermal spectroscopy for gas detection with an anti-resonant hollow-core optical fiber *Nat. Commun.* **11** 847
[108] Chen W, Qiao S, Zhao Z, Gao S, Wang Y and Ma Y 2024 Sensitive carbon monoxide detection based on laser absorption spectroscopy with hollow-core antiresonant fiber *Microw. Opt. Technol. Lett.* **66** e33780
[109] Britto E C, Nizar S M and Krishnan P 2022 A highly sensitive photonic crystal fiber gas sensor for the detection of sulfur dioxide *Silicon* **14** 12665–74
[110] Kumar V, Raghuwanshi S K and Kumar S 2022 Advances in nanocomposite thin-film-based optical fiber sensors for environmental health monitoring—a review *IEEE Sens. J.* **22** 14696–707
[111] Ernst A and Zibrak J D 1998 Carbon monoxide poisoning *New Engl. J. Med.* **339** 1603–8
[112] Cassia R, Nocioni M, Correa-Aragunde N and Lamattina L 2018 Climate change and the impact of greenhouse gasses: CO_2 and NO, friends and foes of plant oxidative stress *Front. Plant Sci.* **9** 273
[113] Zhao Y, Liu Y, Han B, Wang M, Wang Q and Zhang Y 2023 Fiber optic volatile organic compound gas sensors: a review *Coord. Chem. Rev.* **493** 215297

[114] Chen C and Feng W 2013 Intensity-modulated carbon monoxide gas sensor based on cerium dioxide-coated thin-core-fiber Mach-Zehnder interferometer *Opt. Laser Technol.* **152** 108183
[115] Raub J A, Mathieu-Nolf M, Hampson N B and Thom S R 2000 Carbon monoxide poisoning—a public health perspective *Toxicology* **145** 1–14
[116] Chen C, Huang X and Feng W 2023 High sensitivity trace carbon monoxide gas sensor based on endlessly single-mode photonic crystal fiber sagnac interferometer *Z. Naturforsch.* A **78** 105–12
[117] Feng W, Deng D, Yang X, Liu W, Wang M, Yuan M, Peng J and Chen R 2018 Trace carbon monoxide gas sensor based on PANI/Co_3O_4/CuO composite membrane-coated thin-core fiber modal interferometer *IEEE Sens J.* **18** 8762–6
[118] Fan H-L *et al* 2024 Ni-MOF-74 film coated optical microfiber CO sensor cascaded with spiral fiber grating temperature compensator *Opt. Fiber Technol.* **82** 103590
[119] Wang L, Zhou J, Chen Y, Xiao L, Huang G, Huang X and Yang X 2021 An intensity modulated fiber-optic carbon monoxide sensor based on Ag/Co-MOF *in situ* coated thin-core fiber *Z. Naturforsch.* A **76** 881–9
[120] Zhou J, Dai J, Yang S, Wang Y and Yang M 2019 Highly sensitive optical fiber sensor of carbon monoxide based on fabry–perot interferometer and gold-based catalysts *Opt. Eng.* **58** 1
[121] Zhang Y, Yu W, Wang D, Zhuo R, Fu M and Zhang X 2022 Carbon monoxide detection based on the carbon nanotube-coated fiber gas sensor *Photonics* **9** 1001
[122] Kabir M, Habiba U E, Khan W, Shah A, Rahim S, Rios-Escalante P R D L, Farooqi Z-U-R, Ali L and Shafiq M 2023 Climate change due to increasing concentration of carbon dioxide and its impacts on environment in 21st century; a mini review *J. King Saud. Univ. Sci.* **35** 102693
[123] Gall E T, Cheung T, Luhung I, Schiavon S and Nazaroff W W 2021 Real-time monitoring of personal exposures to carbon dioxide *Build Environ.* **104** 59–67
[124] Wolfbeis O S, Kovács B, Goswami K and Klainer S M 1998 Fiber-optic fluorescence carbon dioxide sensor for environmental monitoring *Mikrochim. Acta.* **129** 181–8
[125] Munkholm C 1988 A fiber-optic sensor for CO_2 measurement *Talanta* **35** 109–12
[126] Wysokiński K, Napierała M, Stańczyk T, Lipiński S and Nasiłowski T 2020 Study on the sensing coating of the optical fibre CO_2 sensor *Sensors* **15** 31888–903
[127] Chu C-S, Lo Y-L and Sung T-W 2011 Review on recent developments of fluorescent oxygen and carbon dioxide optical fiber sensors *Photonic Sens.* **1** 234–50
[128] Martan T, Mares D and Prajzler V 2023 Local detection of gaseous carbon dioxide using optical fibers and fiber tapers of single-cell dimensions *Sens. Actuators* B **375** 132887
[129] Chu C-S and Lo Y-L 2009 Highly sensitive and linear optical fiber carbon dioxide sensor based on sol–gel matrix doped with silica particles and HPTS *Sens. Actuators* B **143** 205–10
[130] Liu L, Morgan S P, Correia R, Lee S-W and Korposh S 2020 Multi-parameter optical fiber sensing of gaseous ammonia and carbon dioxide *J. Lightwave Technol.* **38** 2037–45
[131] Wu J, Yin M, Seefeldt K, Dani A, Guterman R, Yuan J, Zhang A P and Tam H-Y 2018 *In situ* μ-printed optical fiber-tip CO2 sensor using a photocrosslinkable poly(ionic liquid) *Sens. Actuators* B **259** 833–9
[132] Kim K-J, Culp J T, Ellis J E and Reeder M D 2022 Real-time monitoring of gas-phase and dissolved CO_2 using a mixed-matrix composite integrated fiber optic sensor for carbon storage application *Environ. Sci. Technol.* **56** 10891–903

[133] Sardar M R, Faisal M and Ahmed K 2020 Design and characterization of rectangular slotted porous core photonic crystal fiber for sensing CO_2 gas *Sens. Biosens. Res.* **30** 100379
[134] Sardar M R, Faisal M and Ahmed K 2021 Simple hollow core photonic crystal fiber for monitoring carbon dioxide gas with very high accuracy *Sens. Biosens. Res.* **31** 100401
[135] de Araujo Silva A, Mijam Barea L A and De Francisco C A 2023 Optical sensor based on an anti-resonant hollow-core fiber for simultaneous detection of methane, carbon monoxide, and nitrogen monoxide: proposal and simulation *J. Opt. Soc. Am.* B **40** C21
[136] Charpentier F *et al* 735610
[137] Dashtban Z, Salehi M R and Abiri E 2022 High sensitivity all-optical sensor for detecting toxic gases using hollow-core photonic crystal fiber composed of magnesium fluoride *Opt. Fiber Technol.* **72** 102958
[138] Pathak A K and Viphavakit C 2022 A review on all-optical fiber-based VOC sensors: heading towards the development of promising technology *Sens. Actuators* A **338**
[139] Sidek O and Bin Afzal M H 2011 A review paper on fiber-optic sensors and application of PDMS materials for enhanced performance *2011 IEEE Symposium on Business, Engineering and Industrial Applications (ISBEIA)* (Piscataway, NJ: IEEE) pp 458–63
[140] Ning X, Yang J, Zhao C L and Chan C C 2016 PDMS-coated fiber volatile organic compounds sensors *Appl. Opt.* **55** 3543
[141] Khan M R R, Kang B-H, Lee S-W, Kim S-H, Yeom S-H, Lee S-H and Kang S-W 2013 Fiber-optic multi-sensor array for detection of low concentration volatile organic compounds *Opt. Express* **21** 20119
[142] Elosua C, Arregui F J, Zamarreño C R, Bariain C, Luquin A, Laguna M and Matias I R 2012 Volatile organic compounds optical fiber sensor based on lossy mode resonances *Sens. Actuators* B **173** 523–9
[143] Aguado C E 2006 Volatile-organic-compound optic fiber sensor using a gold-silver vapochromic complex *Opt. Eng.* **45** 044401
[144] Alimagham F, Platkov M, Prestage J, Basov S, Izakson G, Katzir A, Elliott S R and Hutter T 2019 Mid-IR evanescent-field fiber sensor with enhanced sensitivity for volatile organic compounds *RSC Adv.* **9** 21186–91
[145] Wang J, Yang J, Zou D, Yang J, Qiao G, Wang H and Wang R 2019 Tailor-made photonic crystal fiber sensor for the selective detection of formaldehyde and benzene *Opt. Fiber Technol.* **52** 101941
[146] Topliss S M, James S W, Davis F, Higson S P J and Tatam R P 2010 Optical fibre long period grating based selective vapour sensing of volatile organic compounds *Sens. Actuators* B **143** 629–34
[147] Sun L *et al* 2017 High sensitivity ammonia gas sensor based on a silica-gel-coated microfiber coupler *J. Light. Technol.* **35** 2864–70
[148] Devendiran S and Sastikumar D 2018 Fiber optic gas sensor based on light detection from the samarium oxide clad modified region *Opt. Fiber Technol.* **46** 215–20
[149] He T, Wang W, He B G and Chen J 2023 Review on optical fiber sensors for hazardous-gas monitoring in mines and tunnels *IEEE Trans. Instrum. Meas.* **72** 7003722
[150] Joe H-E, Zhou F, Yun S-T and Jun M B G 2020 Detection and quantification of underground CO2 leakage into the soil using a fiber-optic sensor *Opt. Fiber Technol.* **60** 102375

[151] García J A, Monzón-Hernández D, Cuevas O, Noriega-Luna B and Bustos E 2016 Optical fiber detector for monitoring volatile hydrocarbons during electrokinetic treatment of polluted soil *J. Chem. Technol. Biotechnol.* **91** 2162–9

[152] Deng Y and Jiang J 2022 Optical fiber sensors in extreme temperature and radiation environments: a review *IEEE Sens. J.* **22** 13811–34

[153] Joe H E, Yun H, Jo S H, Jun M B G and Min B K 2018 A review on optical fiber sensors for environmental monitoring *Int. J. Precis. Eng. Manuf.—Green Technol.* **5** 173–91

[154] Li J and Zhang M 2022 Physics and applications of raman distributed optical fiber sensing *Light. Sci. Appl.* **11** 128

[155] Yuan Q, Chai J, Zhang D, Liu J, Li Y and Yin G 2020 Monitoring and characterization of mining-induced overburden deformation in physical modeling with distributed optical fiber sensing technology *J. Light. Technol.* **38** 881–8

[156] Elosua C, Matias I, Bariain C and Arregui F 2006 Volatile organic compound optical fiber sensors: a review *Sensors* **6** 1440–65

[157] Zhan Z, Cantono M, Kamalov V, Mecozzi A, Müller R, Yin S and Castellanos J C 2021 Optical polarization–based seismic and water wave sensing on transoceanic cables *Science (1979)* **371** 931–6

[158] Lindsey N J, Dawe T C and Ajo-Franklin J B 2019 Illuminating seafloor faults and ocean dynamics with dark fiber distributed acoustic sensing *Science (1979)* **366** 1103–7

[159] Gao H, Jiang Y, Cui Y, Zhang L, Jia J and Jiang L 2018 Investigation on the thermo-optic coefficient of silica fiber within a wide temperature range *J. Light. Technol.* **36** 5881–6

[160] Clowes J R, McInnes J, Zervas M N and Payne D N 1998 Effects of high temperature and pressure on silica optical fiber sensors *IEEE Photonics Technol. Lett.* **10** 403–5

[161] Lemaire P J and Lindholm E A 2007 Hermetic optical fibers *Specialty Optical Fibers Handbook* (Amsterdam: Elsevier) pp 453–90

[162] Shao Z, Wu Y, Wang S, Wang Y, Sun Z, Wang W, Liu Z and Liu B 2022 All-sapphire fiber-optic pressure sensors for extreme harsh environments *Opt. Express* **30** 3665

[163] Huang C, Lee D, Dai J, Xie W and Yang M 2015 Fabrication of high-temperature temperature sensor based on dielectric multilayer film on sapphire fiber tip *Sens. Actuators A* **232** 99–102

[164] Ruiz-Lombera R, Laarossi I, Rodriguez-Cobo L, Quintela M A, Lopez-Higuera J M and Mirapeix J 2017 distributed high-temperature optical fiber sensor based on a brillouin optical time domain analyzer and multimode gold-coated fiber *IEEE Sens. J.* **17** 2393–7

[165] Zhang Y, Yuan L, Lan X, Kaur A, Huang J and Xiao H 2013 High-temperature fiber-optic fabry–perot interferometric pressure sensor fabricated by femtosecond laser *Opt. Lett.* **38** 4609

[166] Wang M *et al* 2020 Multiplexable high-temperature stable and low-loss intrinsic fabry-perot in-fiber sensors through nanograting engineering *Opt. Express* **28** 20225

[167] Marcon L, Chiuchiolo A, Castaldo B, Bajas H, Galtarossa A, Bajko M and Palmieri L 2022 The characterization of optical fibers for distributed cryogenic temperature monitoring *Sensors* **22** 4009

[168] Lu X, Soto M A and Thevenaz L 2018 Impact of the fiber coating on the temperature response of distributed optical fiber sensors at cryogenic ranges *J. Light. Technol.* **36** 961–7

[169] Girard S *et al* 2019 Overview of radiation induced point defects in silica-based optical fibers *Rev. Phys.* **4** 100032

[170] Lyu D, Peng J, Huang Q, Zheng W, Xiong L and Yang M 2021 Radiation-resistant optical fiber fabry-perot interferometer used for high-temperature sensing *IEEE Sens. J.* **21** 57–61
[171] Girard S, Kuhnhenn J, Gusarov A, Brichard B, Van Uffelen M, Ouerdane Y, Boukenter A and Marcandella C 2013 Radiation effects on silica-based optical fibers: recent advances and future challenges *IEEE Trans. Nucl. Sci.* **60** 2015–36
[172] Baldwin C S 2014 Brief history of fiber optic sensing in the oil field industry *Proc. Fiber Optic Sensors and Applications XI***vol 9098** ed H H Du, G Pickrell, E Udd, C S Baldwin, J J Benterou and A Wang (SPIE) p 909803
[173] Ashry I, Mao Y, Wang B, Hveding F, Bukhamsin A, Ng T K and Ooi B S 2022 A review of distributed fiber–optic sensing in the oil and gas industry *J. Light. Technol.* **40** 1407–31
[174] Rong Q and Qiao X 2019 FBG for oil and gas exploration *J. Light. Technol.* **37** 2502–15
[175] Edouard M N, Okere C J, Dong P, Ejike C E, Emmanuel N N and Muchiri N D 2022 Application of fiber optics in oil and gas field development—a review *Arab. J. Geosci.* **15** 539

IOP Publishing

Glass-based Materials
Advances in energy, environment and health
Sathish-Kumar Kamaraj and Arun Thirumurugan

Chapter 6

Materials based on glasses for energy generation in electrochemical systems

Omar Francisco González Vázquez and José de Jesús Serralta Macías

Starting from the fact that glasses are amorphous ceramic materials produced by the rapid cooling of a liquid is not sufficient to fully define them. This is because a series of characteristics such as their composition, synthesis temperatures, cooling rates, and the addition of intermediate or modifying compounds lead to the generation of a myriad of vitreous materials with different properties. The engineering properties of glasses are incredibly diverse and surpass the performance of many other commonly used materials. For example, glasses are much harder than many untreated metals, with a tensile strength ranging from 24 to 69 MPa. They exhibit high ductility, low coefficient of expansion, and thermal conductivity. Moreover, glasses are highly resistant to the attack of reactive chemicals and can withstand high temperatures, up to around 1580 °C [1]. These extraordinary properties have enabled remarkable technological advancements and broadened their application in various sectors, including the food industry, automotive industry, construction, electronics, pharmaceuticals, decoration, and energy production. Among these sectors, energy production has shown particular interest in harnessing the diverse properties of these materials.

It is well-known that energy production from fossil fuels reached its peak during the 20th century, leading to unprecedented development across all sectors of human history. However, it also resulted in increased air, soil, and water pollution, giving rise to a host of social, health, and environmental problems. Furthermore, the global population demands an ever-increasing energy supply, while reserves of these fuels are steadily declining [2]. This poses a significant challenge that must be addressed promptly, where emerging technologies play a crucial role in creating new clean, affordable, and efficient energy production. In this context, scientists have focused their attention on materials such as glass, which offer design flexibility, allowing for a variety of properties. Consequently, glass can be utilized in diverse ways for energy production, storage, and transportation.

doi:10.1088/978-0-7503-5904-7ch6

There are primarily three established environmentally friendly energy production technologies that utilize glasses as fundamental materials: solar energy, which harnesses the optimized optical properties of glasses; wind energy, which utilizes the mechanical properties of glass as reinforcement in the form of fibers; and nuclear energy, which leverages the chemical stability of glasses to store and contain nuclear fuels [3], However, there are unconventional energy production technologies based on electrochemical systems that promise to be superior to the already established ones and could also incorporate vitreous materials in their manufacturing processes. For instance, the production of hydrogen in electrochemical cells, where exchange membranes are made of glass; or the development of chemical batteries with vitreous electrolytes; as well as the use of glass seals in solid oxide fuel cells, to name a few.

This chapter will review the properties, design, and use of vitreous materials in electrochemical technologies for energy production, including an overview of trends in glass synthesis.

6.1 A brief introduction to electrochemical systems for energy production

Electrochemical systems for energy production involve redox reactions of certain chemical agents, resulting in the production of new agents and a usable electric current. This production of electrical energy must occur in a single step; in other words, it cannot go through intermediate stages, such as converting chemical energy to mechanical energy and then to electrical energy. Electrochemical systems that enable the direct production of electrical energy in a single step are referred to as fuel cells. While there are more electrochemical systems related to energy, only fuel cells can 'generate' energy autonomously. Electrochemical systems like batteries, supercapacitors, and electrolyzers do not produce energy per se. Many substances that, upon decomposition, release electrically charged by-products, both negatively and positively, can be used as combustible substances.

Fuel cells consist of three basic elements: an anode, a cathode, and an electrolyte positioned between the first two. However, they often include a fourth component, which is a membrane separating the anodic and cathodic reactions (figure 6.1) [4]. In this context, the efficiency of energy production is the key to success in fuel cell design, demanding meticulous material selection to make these systems as efficient as possible. Efficiency is understood as the power density generated concerning the area or volume of the system, or as the conversion of a reactant into electrical energy [5]. A crucial parameter for achieving high efficiency in fuel cells is the reduction of internal resistance, caused by the opposition of the conduction materials, electrodes, and membrane to the transport of electrons and protons. This underscores the importance of careful material selection for their construction. Additionally, it is imperative to understand that oftentimes, the electrolytes or substances interacting with the cell components are highly corrosive due to their acidic or alkaline nature or are very strong solvents. Consequently, container materials or separators must have the capacity to resist these substances. Another parameter directly impacting the efficiency of these systems is particle selectivity. In two-chamber fuel cells, the

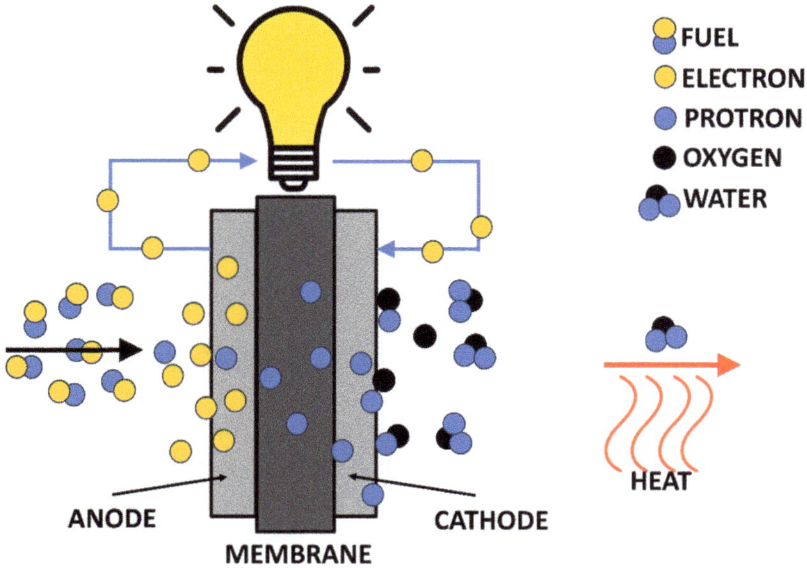

Figure 6.1. Generic operation of a two-chamber fuel cell.

semipermeable membrane must be highly selective, allowing only the passage of protons. As evident, the requirements for fuel cells are highly specific, and vitreous materials have the potential to meet several of these criteria.

6.2 Properties of glasses for their application in electrochemical cells

As mentioned earlier, the construction of fuel cells demands highly specific characteristics to achieve optimal performance, aiming to deliver the highest power density or the highest degree of conversion of fuel into electrical energy. All these properties can be categorized into three major groups: structural, chemical, and ionic conductivity. Each of these groups encompasses several sub-properties and is intertwined with the aspects mentioned in the previous section.

6.2.1 Structural properties

Glasses, being amorphous materials, often possess a certain level of natural porosity. This porosity allows the formation of channels within their structure, granting them a degree of permeability. Textural properties such as volumetric porosity, average pore diameter, specific surface area, coefficient of structural resistance, and tortuosity coefficient are crucial in achieving specific permeability and load resistance. Pore sizes, in this context, play a fundamental role in achieving good permeability and, consequently, electrical conductivities that glasses inherently lack, given their electrical insulating nature [6]. Porous glasses are essential for constructing semipermeable membranes, and each of their structural characteristics is vital for ensuring that the membrane conducts protons from the anodic chamber to the cathodic one.

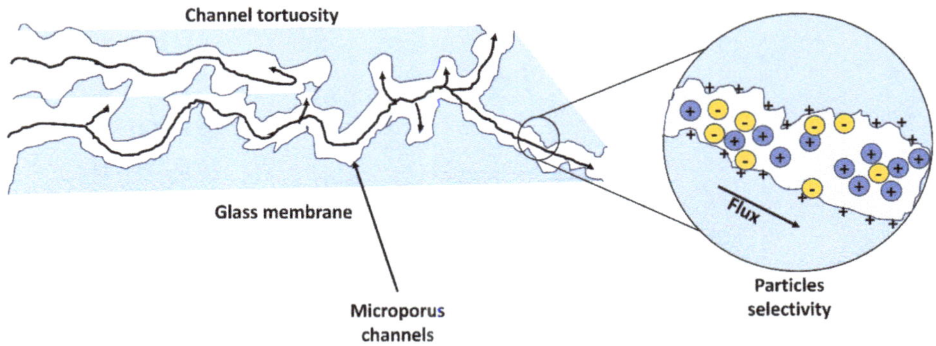

Figure 6.2. Principal properties of microporous glass membranes for fuel cells.

Microporous glasses (those with pores smaller than 2 nm) are typically the most effective materials for selecting particles that are not desired to pass through the glass membranes. The interfacial contact between the carrier electrolyte and the glass membrane surface is crucial for selectivity (figure 6.2) [7]. It has even been proven that the reduced size of the microchannels can help rectify the flow of the carrier electrolyte as it passes through the glass membranes, owing to the glass's surface composition [8], because the surface charges lead to a slowing down of the flow, allowing intimate contact between undesired particles and the glass surface, thereby achieving superior selectivity.

In the same vein, the tortuosity of the electrolyte pathway ensures that the residence time is appropriate for adsorbing particles that should not reach the cathodic chamber. Taken together, all these structural properties enable high selectivities, good permeabilities, and, consequently, improved ionic conductivities.

6.2.2 Chemical properties

As seen before, the composition of glasses used as membranes in fuel cells is crucial for the selectivity in the flow of charged particles. This composition is intimately linked to their synthesis method since the selective substances can be part of the glass matrix or added as decoration, doping, or fixation particles. These compounds, besides providing selectivity, allow for acceptable coulombic efficiencies. Materials such as graphene oxides or carbon nanoparticles have proven to be a viable option for this purpose [9]. Additionally, these functionalizing agents must be resistant to the corrosion and reactivity of the electrolytes to which they will be exposed. Glasses are typically chosen to meet this requirement, making materials as robust as graphene-based materials a suitable choice.

In this regard, polymeric materials offer higher performance than glass membranes, as they are often resistant, inert, and easily functionalizable. However, the cost of commercial membranes like NAFION is prohibitively high, making their large-scale use unfeasible. Consequently, various researchers have chosen to create hybrid glass and polymer membranes. These hybrids help reduce production costs while maintaining good performance in terms of chemical resistance, selectivity, and pore size control [10].

The concept of 'anionic subnetwork' in glasses refers to the internal structure of the glass where negatively charged ions (anions) are arranged in a three-dimensional network. Glasses are amorphous materials, meaning they lack a regular and ordered crystalline structure. Instead, their atoms or ions are arranged in a disorderly way, granting them specific properties.

In the context of glasses, especially in ionic glasses, anions can be part of this three-dimensional amorphous structure. A common example is silica glass (quartz glass), which mainly consists of silicon (Si) and oxygen (O) atoms. In this glass, oxygen ions (O^{2-}) form a three-dimensional network within the vitreous structure.

Different types of ions can be introduced into ionic glasses to alter their electrical and chemical properties. For instance, in conductive ionic glasses, lithium ions (Li^+) or sodium ions (Na^+) are introduced into the glass's anionic network. These mobile ions can move within the glass structure, allowing ionic conductivity and making these glasses useful in applications such as solid-state battery electrolytes.

The proper selection of constituents in the anionic subnetwork of glasses can affect the redox stability of the glass, which, in turn, can impact its ability to act as an electrolyte. Moreover, the ionic conductivity of these glasses depends on the concentration and mobility of ions within the glass structure, which, in turn, rely on the glass's chemical composition. Generally, manipulating the chemical composition of ionically conductive glasses allows for the adjustment of their electrochemical properties to suit various applications in electrochemical devices [11].

On the other hand, there are studies on glass membranes doped with ionic liquids that enable higher ionic conductivity. This allows the glass to serve merely as a support, with the ionic liquid carrying out the exchange work [12].

Chemical properties also influence physical properties, such as resistance to relatively high operating temperatures. It is well-known that at elevated temperatures, both ionic and electrical conductivity tend to decrease. This is one of the key reasons why glasses are chosen for use in fuel cells, as they maintain their conductivity even under high-temperature conditions.

6.2.3 Conductivity properties

As mentioned in previous sections, one of the primary characteristics to consider in glasses used as membranes in fuel cells is their electrical or ionic conductivity. Specifically, when discussing separation membrane conductivity, it pertains to ionic conductivity since glasses are known for their high intrinsic electrical resistance. Ionic conductivity in glasses refers to the ability of ions within the glass to carry electric charge through the material. Ionic glasses are amorphous solid materials exhibiting significant ionic conductivity at room temperature or moderately elevated temperatures. This ionic conductivity is due to the presence of mobile ions in the glass structure. Ionic conductivity in glasses depends on various factors, including the concentration and mobility of ions in the vitreous structure, as well as the material's temperature. At higher temperatures, ions possess more kinetic energy and are more likely to move, generally increasing the glass's ionic conductivity. In this context, achieving good ionic conductivity is associated with the glass's ability to adsorb water and facilitate ion transport [13]. It has been reported that ionic conductivity can be improved in glasses when the relative humidity reaches 85%. Additionally, membranes doped with acids such as phosphomolybdic acid (PMA) or phosphotungstic acid (PWA) have been shown to enhance ionic conductivity characteristics [12, 14].

It is accurate to say that ionic conductivity capabilities are primarily attributed to the chemical composition of glasses, as well as to a well-defined structural arrangement that allows the membrane to have high membrane density [10].

6.3 Use of glasses as seals and solid electrolytes in fuel cells

A solid oxide fuel cell is a type of fuel cell that utilizes a solid electrolyte to transport oxygen ions from the cathode to the anode. These cells operate at high temperatures (between 500 °C and 1000 °C) and can use a variety of fuels such as hydrogen, methane, and carbon monoxide to produce electricity. Solid oxide fuel cells have the advantage of being highly efficient and having low pollutant emissions. These cells require glass seals to seal the cell components and prevent leaks of fuel gas. Additionally, glass seals help maintain the structural integrity of the cell and prevent cross-contamination between gases used in the cell.

A glass seal is a layer of glass used to seal components of a solid oxide fuel cell. The glass seal is placed between the fuel cell components to prevent gas leaks and ensure that the components are properly sealed. The glass seal must possess a combination of thermal, chemical, mechanical, and electrical properties to guarantee its long-term stability and resistance to thermal cycling. Moreover, the glass seal must withstand gas pressure, vibration, and thermal cycles during the fuel cell's operation [15]. The thermal and chemical properties of glasses are highly specific for this application. The challenge lies in designing seals that are thermally stable and durable to prevent replacements and increase the maintenance cost of fuel cells. It is crucial to prevent interfacial reactions and minimize thermal expansion [16]. In this case, due to the elevated temperatures involved, properties such as softening temperature and glass transition temperature need to be carefully considered. Therefore, thermal stability is of paramount importance. The thermal stability of a glass depends on the difference between its melting temperature and its glass transition temperature. The greater this difference (i.e., the wider the supercooling range), the higher the thermal stability of the glass, making it more likely to form a stable glass [17].

Although glasses offer several advantages as sealing materials in solid oxide fuel cells, they also come with disadvantages. Firstly, glasses are brittle and can break or crack under mechanical or thermal loads. This can be problematic in fuel cells operating at high temperatures and subjected to repeated thermal cycles. This challenge has led to the development of hybrid materials combining glasses and ceramics. These compounds can exhibit better deformation resistance and increased durability in aggressive environments, such as those found in solid oxide fuel cells. Additionally, composite seals can allow a greater degree of relative movement between the bonding surfaces without causing leaks, which can be advantageous in specific applications [18].

On another note, glass can also be utilized as a solid electrolyte. The primary distinction between solid electrolytes and selective porous glass membranes lies in the fact that solid glass electrolytes can conduct electricity, primarily through various ions, especially alkali ions, enabling current conduction. In contrast, in glass membranes serving as selective barriers, ion conduction occurs through diffusion [19].

The primary advantage of using glass as a solid electrolyte is its high ionic conductivity, wide electrochemical window, good chemical and thermal stability, and the ability to be manufactured in complex shapes and sizes. Moreover, glass solid electrolytes can be safer and less flammable than liquid electrolytes used in some cells or batteries [20].

6.4 Design and manufacturing of electroselective membranes for use in fuel cells

In the relentless pursuit of more efficient and sustainable energy solutions, the design and manufacturing of electroselective membranes for implementation in fuel cells have become a fundamental research area. These membranes play a crucial role in separating and facilitating the selective transfer of ions, enabling the highly efficient direct conversion of chemical energy into electricity. The intersection of materials engineering, nanotechnology, and electrochemistry has led to significant advances in the design and production of these electroselective membranes, used in various applications from hydrogen fuel cells to renewable energy devices. This multidisciplinary approach has not only enhanced the efficiency and durability of fuel cells but has also paved the way for a cleaner and more sustainable energy future. In this context, this study delves into the fascinating world of designing and manufacturing these membranes, exploring the latest developments, challenges, and applications in the field of fuel cell technologies.

6.4.1 Synthesis methods

There are several methods for manufacturing glass membranes for use in fuel cells, each with various advantages and disadvantages. The adaptability of the process depends on the desired product and the available infrastructure. Below, we outline the most commonly used methods for this purpose.

6.4.1.1 Sol–gel method

The sol–gel method is a chemical technique used to produce inorganic materials such as ceramics and glasses from liquid solutions. The process involves the hydrolysis of inorganic precursors to form a colloidal solution (sol), which then undergoes polycondensation to create a three-dimensional network (gel). The resulting gel is dried and calcined to produce a solid material (figure 6.3) [21]. This process allows for controlling the material density based on the precursor concentration and drying conditions.

Some studies have synthesized glass materials for fuel cell membranes using this method, finding significant advantages in conductivity and ease of manufacturing.

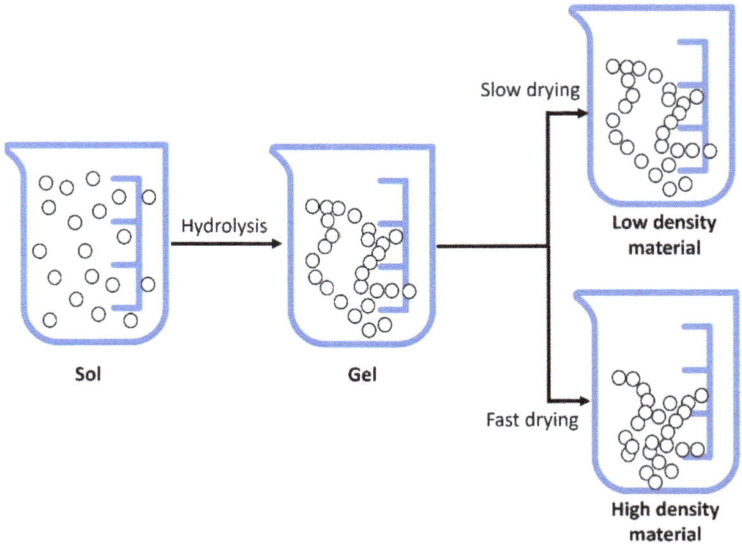

Figure 6.3. Sol–gel process for solid material synthesis.

Nogami *et al* mentioned the sol–gel synthesis used to prepare binary P_2O_5–SiO_2 glass. The process involved the reaction of $PO(OCH_3)_3$ with $Si(OC_2H_5)_4$, followed by additional hydrolysis with H_2O and $CHONH_2$. The resulting solution was allowed to stand for about a month to form a rigid gel with incorporated porosity. The gel was heated to remove water and solvents, contracting due to the dehydration-condensation of hydroxyl groups. The porous structure of the gel remained unchanged in the resulting glass after heating it to temperatures below 900 °C, where the synthesized glass exhibited high proton conductivity of 2.2×10^{-2} S cm^{-1} at 50 °C, making it suitable as an electrolyte in fuel cells [21].

Uma *et al* developed a synthesis process to obtain a P_2O_5–TiO_2–SiO_2 glass membrane using the sol–gel method. They formed a suspension with appropriate quantities of $Ti(OC_4H_9)_4$, $Si(OC_2H_5)_4$, and $PO(OCH_3)_3$, which was then partially hydrolyzed with a solution of H_2O (such as 0.15 N HCl) and C_2H_5OH in molar ratios of 1:1 per mol of TEOS (tetraethyl orthosilicate) with constant stirring. The material achieved conductivities in the range of 10^{-4}–10^{-2} S cm^{-1} at 70% relative humidity and within the temperature range of 30 °C–90 °C. Conductivity values of 1.3×10^{-2} S cm^{-1} were achieved at 90 °C, with materials thermally stable up to 300 °C and chemically stable with H_2O during prolonged exposures [22].

6.4.1.2 Chemical vapor deposition (CVD) method

Chemical vapor deposition (CVD) is a process used to manufacture thin films of solid materials on substrates. In the CVD process, gaseous chemical precursors are introduced into a high-temperature reaction chamber, where they react to form a solid material that deposits onto a substrate. The process involves several steps. First, the reaction chamber is prepared, the substrate is placed inside, and all gases unrelated to the process are evacuated. Then, precursor chemicals in the form of gases are introduced into the reaction chamber. Inside the chamber, these precursors decompose and react, generating solid products on the substrate's surface. This entire process is meticulously controlled, including temperature, pressure, and precursor concentration. Finally, the material is cooled and removed from the chamber. This method allows for precise control over the size of the deposited material film, enabling the production of very thin materials [23].

Prakash *et al* synthesized a phosphorus-doped silicon dioxide material (PSG) to be used as a thin membrane in fuel cells. The synthesis process involved the deposition of thin films of this material onto a metalized substrate through the reaction of silane, phosphine, and nitrous oxide in a plasma-enhanced chemical vapor deposition (PECVD) system. A systematic variation of critical deposition parameters (substrate temperature, RF power, pressure, and gas composition) was investigated. Ionic conductivities of up to 2.54×10^{-4} S cm^{-1} were achieved for PSG films, which is 250 times higher than the conductivity of pure SiO_2 films (1×10^{-6} S cm^{-1}) under identical deposition conditions [24].

Furthermore, Li *et al* synthesized composite glass membranes coated with Nafion for use in microfabricated fuel cells. A Nafion solution containing less than 10% of perfluorosulfonic copolymer/PTFE resin manufactured by DuPont and diethylene glycol in a 5:1 ratio was used. This solution was poured onto the surface of phosphorus-doped silicon dioxide after treatment with an adhesion promoter. Phosphorus-doped silicon dioxide glass was conducted in a PECVD system, Unaxis, at temperatures ranging from 100 °C to 250 °C using a mixture of nitrous oxide and silane/phosphine. The conductivity of PSG deposited by PECVD was nearly two orders of magnitude higher compared to undoped SiO_2, allowing the membrane to be thicker to enhance mechanical strength without compromising cell performance through increased cell resistance [25].

6.4.1.3 Sintering method

The sintering of glasses involves the manufacture of vitreous materials through controlled processes of melting and cooling. There are various techniques for synthesizing glasses, and the choice of method depends on the specific application and the type of glass desired. However, the general process consists of selecting the precursors, primarily oxides such as silicon oxide, sodium, aluminum, and/or calcium. Care is taken to ensure that these materials have minimal or no impurities. Subsequently, a mixture is prepared in desired proportions, according to the desired formula. This mixture is melted at high temperatures, and during this process, the components chemically combine, form, or mold, and then cool suddenly or slowly, depending on the desired degree of crystallinity. This process is known as sintering [26].

Zeng *et al* developed the synthesis of proton-exchange membranes composed of functionalized sintered mesoporous silica nanocomposites with anhydrous phosphoric acid for fuel cells. In this process, Pluronic P123 surfactant and hydrochloric acid were dissolved in deionized water. After complete dissolution, butanol was added, and after stirring for 1 h at 45 °C, tetraethyl orthosilicate (TEOS) was introduced. The mixture was vigorously stirred at 45 °C for 24 h and then aged at 100 °C for 24 h. The mixture was dried at room temperature. The resulting powder was calcined at 650 °C for 6 h to remove the template, using N_2 as a protective gas at temperatures below 200 °C. The prepared mesoporous silica powder was pressed into a matrix to form mesoporous silica discs, followed by sintering in a temperature range of 550 °C to 850 °C. The PA-mesoporous silica membrane and the PA-porous silica membrane exhibited a conductivity of approximately 0.1 S cm^{-1} at 190 °C without external humidification control. This conductivity was superior without humidification compared to Nafion 117 and PBI/PA membranes [27].

The same author also investigated the synthesis of an inorganic proton-exchange membrane based on sintered mesoporous silica and phosphoric acid for use in high-temperature proton-exchange membrane fuel cells. The process began with the synthesis of mesoporous silica powder with a cubic and bicontinuous structure using a triblock copolymer P123 as a template under acidic conditions. Subsequently, ordered mesoporous silica powder was obtained by removing the template through heat treatment. The membrane was sintered at a high temperature (650 °C) to form a dense membrane with high porosity and ordered mesoporous structure. The pre-sintering of mesoporous silica powder also avoids the drawbacks of hot-pressed mesoporous silica-based PEM membranes. The sintered mesoporous silica membrane was impregnated with phosphoric acid by immersion in an 83% phosphoric acid solution for 24 h. The conductivity achieved by the PA/meso-silica membrane exhibits superior performance compared to PBI/PA membranes, with a proton conductivity measured in the absence of external humidification ranging from 4×10^{-3} to 6×10^{-2} S cm^{-1} in the temperature range of 80 °C–225 °C [28].

6.5 Evaluation of the economic viability of glass membranes

6.5.1 Supplies

The sol–gel process is a widely recognized method for the synthesis of pure glasses [29], binary glasses [30], glass-based composites [31–34], and hybrid polymer-glass membranes [10]. This technique presents an efficient route for fabricating nanoporous and mesoporous membranes or structures at low temperatures, eliminating the need for high-cost equipment. Moreover, materials derived from the sol–gel process allow for control over aspects such as composition, morphology, pore size, specific surface area, and the dispersion of particles within the matrix [35] (figure 6.4).

A typical sol–gel procedure to obtain glass primarily involves combining precursors like tetraethyl orthosilicate (TEOS) [35–40] for SiO_2 synthesis, and either triethylphosphate (TEP) or phosphoric acid (H_3PO_4) for P_2O_5 production. These mixtures are typically prepared at room temperature, introducing solvents (such as water, ethanol, or methanol) and acids (like HCl or NH_3 solution) to regulate the pH within a 1.5–4.5 range [41]. That promotes the hydrolysis and condensation reactions leading to the formation of a colloidal suspension (sol) or a precipitate. At this step, polymeric resins can be introduced to the sol to fabricate a hybrid polymeric glass solution [42]. If a precipitate forms, it undergoes filtration and calcination to yield mesoporous silica powder. Conversely, if a sol is formed, it can be cast or spin-coated onto a substrate or molded. As the solvent is gradually eliminated during the aging phase, a gel structure emerges. This gel then undergoes a

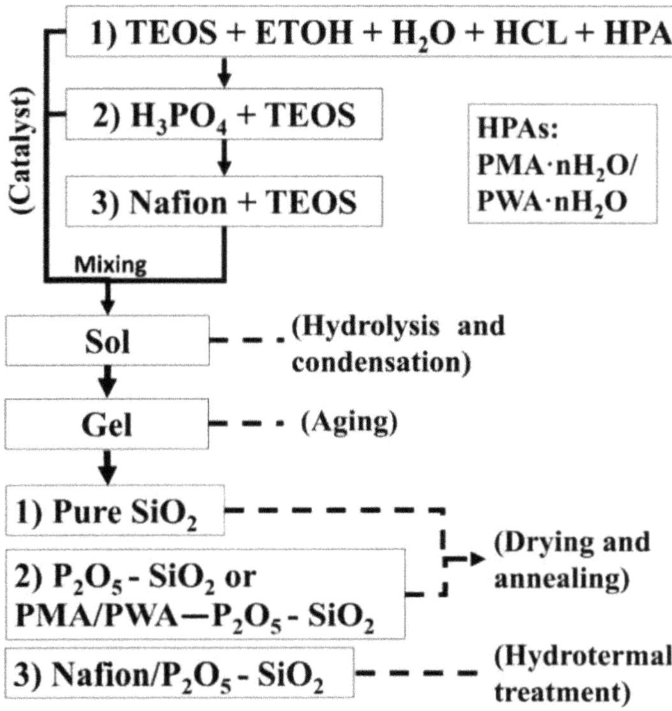

Figure 6.4. Flow chart of glass and glass-based membranes synthesis by sol–gel method.

drying procedure to produce a dried gel (xerogel). If needed, it can be subjected to further heat treatment processes, such as hydrothermal or annealing treatments [42, 43], to eradicate residual organic elements and augment the material's density, culminating in the creation of the ultimate glass membrane.

6.6 Efficiency performance in glass fuel cells

In pure porous SiO_2 glass membranes, protons migrate from the anode to the cathode via surface-adsorbed water and condensed water within the pores [37]. This phenomenon results in proton conductivity, as shown in (figure 6.5). Therefore, the glass membrane is emerging as a promising proton-exchange membrane (PEM) due to its high thermal and chemical stability compared with that of typical polymeric membranes. Particularly, proton-exchange membrane fuel cells (PEMFCs) have garnered significant attention as advanced power sources for both stationary and vehicular systems due to their environmental benefits and superior energy efficiency.

Nowadays, one of the most utilized electrolyte membranes in PEMFCs is the perfluorosulfonic acid (e.g., Dupont's Nafion®). This is mostly attributed to its notable proton conductivity and high chemical stability [42]. Consequently, it's often assumed that PEMFCs mostly utilize polymeric electrolyte membranes. However, innovative PEMs have been developed to substitute polymeric membranes. The above is because, despite the high proton conductivity of polymeric membranes, like

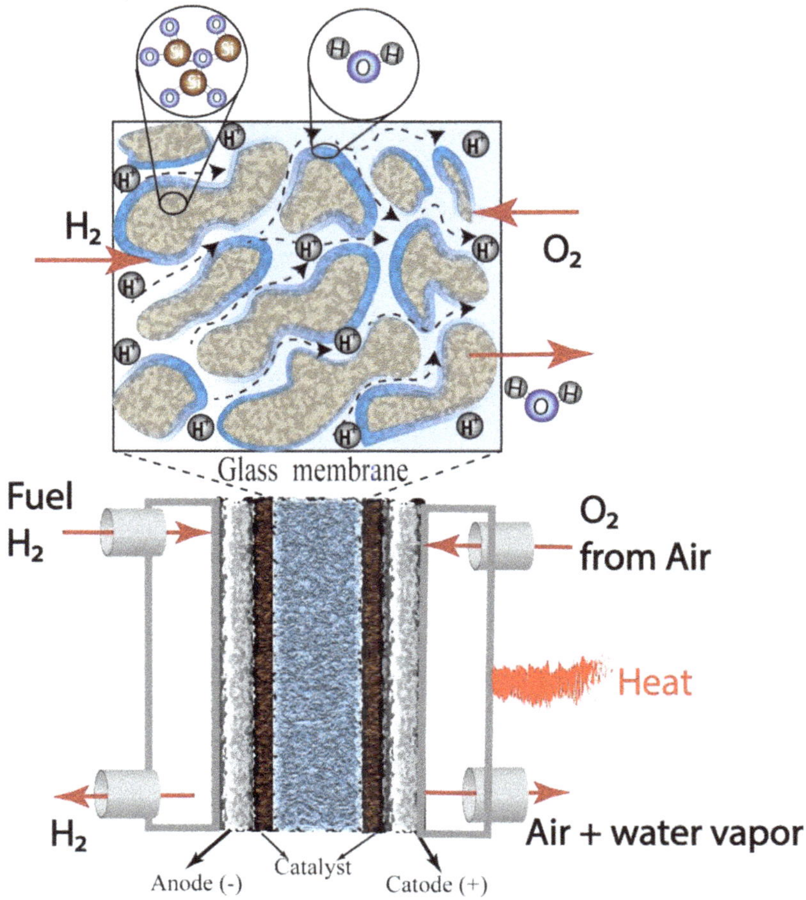

Figure 6.5. Water-assisted proton transport through porous silica glass membrane in PEMFCs.

Nafion®, their application is restricted to operation temperatures below 100 °C due to thermal limitations. Moreover, the high cost of Nafion® membranes coupled with their tendency to swell from water absorption negatively impacts their fuel-cell efficiency [44]. According to several authors, PEMs are grouped into three main categories: (i) polymeric, (ii) ceramic, and (iii) composite membranes [44].

Most inorganic membranes in that category as composed by sol–gel derived proton conductors such as glasses (SiO_2, P_2O_5–SiO_2), ceramics (ZrO_2, Al_2O_3, TiO_2 [42]), and ceramic-glass composites. Notably, despite their classification under ceramics, glasses have distinct properties and performance [44]. Particularly, ceramic PEMs, comprising inorganic, non-metallic crystalline structures, are recognized for their hardness, durability, and thermal stability. However, it's necessary to distinguish ceramics for PEMs, which typically operate at moderately high temperatures, from those in solid oxide fuel cells (SOFCs), such as yttria-stabilized zirconia (YSZ) that facilitate oxygen ion conduction at high temperatures ranging from 600 °C to 1000 °C [45]. On the other hand, glass PEMs, which are amorphous solids, excel in thermal and chemical stability, especially in elevated operational temperatures of 100 °C–200 °C [10, 31]. Nonetheless, the inherent brittleness of pure porous glass membranes and limited proton conductivity ranging from 10^{-6} to 10^{-3} S cm^{-1} render them less favorable for PEM applications [29]. To address this, silicon has been modified with specific acids to improve proton conductivity at low humidity, yielding power densities of 35–85 mW cm^{-2} [13, 46, 47]. The influence of ceramics on glass compaunds such as TiO_2–P_2O_5–SiO_2 [35], ZnO_2–P_2O_5–SiO_2 [39], SiO_2–TiO_2–ZrO_2–P_2O_5, and SiO_2–TiO_2–ZrO_2–P_2O_5–Bi_2O_3 [38] lead to membranes thermally stable to at least 400 °C and chemically stable with H_2O for prolonged exposures. However, those systems showed low proton conductivity 10^{-5}–10^{-3} S cm^{-1} and power density 0.9–2.5 mW cm^{-2} [32, 35, 38]. Generally, the acidification of glass composites with PA, PMA, and PWA has shown a significant improvement in proton conductivity and power density in glass-based membranes. Thus, acid-functionalized ceramic-glass composite such as PMA/ZrO_2–P_2O_5–SiO_2 has been developed that showed 32 mW cm^{-2} [34].

Typically, SiO_2 glass membranes are obtained by the sol–gel method due to low cost and high flexibility in the modification of the chemical composition, dimension, and pore microstructure of membranes. For instance, SiO_2 glass membranes have been combined with other glasses, acids, sulfonates, or a combination of both to obtain binary glass (P_2O_5–SiO_2), acid-functionalized glass (PWA-SiO_2), diethyl methylammonium trifluoromethanesulfonate, [dema][TfO]/SiO_2 hybrid glass [10], and heteroacid-modified binary glass (PWA/PMA-P_2O_5–SiO_2 [13]), respectively. Furthermore, glasses can be combined with ceramics and polymeric materials to obtain composite materials with special characteristics that individual materials lack. In that way, glass and glass-based materials can reach high proton conductivity values of about 10^{-2}–10^{-1} S·cm [30, 31, 42], which are comparable to that of Nafion® and are thus adequate for fuel-cell applications. Table 6.1 summarizes the specific surface area, pore size, conductivity, and power density of several glass and glass-based electrolyte membranes applied to H_2/O_2 fuel cells.

Table 6.1. Specific surface area, pore size, conductivity, and power density of glass and glass-based electrolyte membranes applied to H_2/O_2 fuel cells.

Type of material	System	Specific surface area, pore size (nm)	Conductivity ($S\ cm^{-1}$)	Power density ($mW\ cm^{-2}$)	Operation conditions	References
Pure glass	SiO_2	700–900 $m^2\ g^{-1}$ 1.6–1.8 nm	From 10^{-6} to 10^{-3}	—	400 °C	[29]
Binary glass	P_2O_5–SiO_2,		2.2×10^{-2}	6	50 °C, 90% RH Thickness (d) = 0.4 mm	[36]
Acid-functionalized glasses	Phosphoric acid (PA)/SiO_2	6.6 to 1100 m2 g−1 3.2 to 4.5 nm	10^{-2}	—	100 °C–120 °C 100% RH	[37]
	PWA/PMA-P_2O_5–SiO	—	1.014	41.5	32 °, 85% RH	[13]
	PWA/PMA-P_2O_5–SiO2	178 m2 g−1 5.333 nm	0.17	35	28 °C, 30%RH *d* = 0.55 mm	[47]
	$CsH_5(PO_4)$2–SiO_2	253 m2 g−1 1–40 nm	0.1	85	210 °C, 30%RH *d* = 0.1 m	[46]
Glass-Ceramic composite	P_2O_5–TiO_2–SiO_2	410 m2 g−1 2.4 nm	3.6×10^{-2}	0.088	90 °C 70% RH *d* = 0.5–1.5 mm	[33]
	ZrO_2–P_2O_5–SiO_2	470 m2 g−1 2.5 nm	24.6×10^{-4}	0.9	30 °C, 30%RH	[33]

	SiO_2–TiO_2–ZrO_2–P_2O_5	—	3.6×10^{-5}	0.03	80 °C $d = 1.1$ mm	[38]
	SiO_2–TiO_2–ZrO_2–P_2O_5–Bi_2O_3	—	4.6×10^{-3}	2.5	150 °C $d = 0.8$ mm	[38]
Acid-functionalized composite	PMA/ZrO_2–P_2O_5–SiO_2	250–375 m2 g−1 2.3–2.4 nm	—	32	29 °C, 30% RH	[34]
Hybrid polymer-glass	[dema][TfO]/SiO_2	—	From 10^{-3} to 10^{-2}	—.	200 °C, 90%RH	[12]
	Nafion®/ P_2O_5–SiO_2 (NPS)	—	10^{-1}	207	70 °C $d = 0.5$ mm	[42]
	NPS/SPEAK	6.9 nm	10^{-3}	322	65 °C, 90%RH Thickness 0.03 mm	[10]

References

[1] Axinte E 2011 Glasses as engineering materials: a review *Mater. Des.* **32** 1717–32
[2] Armaroli N and Balzani V 2007 The future of energy supply: challenges and opportunities *Angew. Chem.—Int. Ed.* **46** 52–66
[3] Brow R K and Schmitt M L 2009 A survey of energy and environmental applications of glass *J. Eur. Ceram. Soc.* **29** 1193–201
[4] Sharaf O Z and Orhan M F 2014 An overview of fuel cell technology: fundamentals and applications *Renew. Sustain. Energy Rev.* **32** 810–53
[5] Boudghene Stambouli A and Traversa E 2002 Fuel cells, an alternative to standard sources of energy *Renew. Sustain. Energy Rev.* **6** 295–304
[6] Ermakova L E, Kuznetsova A S, Antropova T V, Volkova A V and Anfimova I N 2020 Structural characteristics and electrical conductivity of porous glasses with different compositions in solutions of sodium, lanthanum and iron(III) chlorides *Colloid J.* **82** 262–74
[7] Altug I and Hair M L 1968 The ion-selective properties of sintered porous glass membranes *J. Phys. Chem.* **72** 2976–81
[8] White H S and Bund A 2008 Ion current rectification at nanopores in glass membranes *Langmuir* **24** 2212–8
[9] Ahilan V, de Barros c c, Bhowmick G D, Ghangrekar M M, Murshed M M, Wilhelm M and Rezwan K 2019 Microbial fuel cell performance of graphitic carbon functionalized porous polysiloxane based ceramic membranes *Bioelectrochemistry* **129** 259–69
[10] Li H, Chen X, Jiang F, Ai M, Di Z and Gu J 2012 Flexible proton-conducting glass-based composite membranes for fuel cell application *J. Power Sources* **199** 61–7
[11] Souquet J L 1981 Electrochemical properties of ionically conductive glasses *Solid State Ion* **5** 77–82
[12] Li H, Jiang F, Di Z and Gu J 2012 Anhydrous proton-conducting glass membranes doped with ionic liquid for intermediate-temperature fuel cells *Electrochim. Acta* **59** 86–90
[13] Uma T and Nogami M 2008 Proton-conducting glass electrolyte *Anal. Chem.* **80** 506–8
[14] Uma T and Nogami M 2007 A novel glass membrane for low-temperature H_2/O_2 fuel cell electrolytes *Fuel Cells* **7** 279–84
[15] Mahapatra M K and Lu K 2010 Seal glass for solid oxide fuel cells *J. Power Sources* **195** 7129–39
[16] Sohn S-B, Choi S-Y, Kim G-H, Song H-S and Kim G-D 2002 Stable sealing glass for planar solid oxide fuel cell *J. Non-Cryst. Solids* **297** 103–12
[17] Fernández-Navarro J-M and Villegas M-A 2013 What is glass? An introduction to the physics and chemistry of silicate glasses *Modern Methods for Analysing Archaeological and Historical Glass, I* (Wiley)
[18] Dev B and Walter M E 2015 Comparative study of the leak characteristics of two ceramic/glass composite seals for solid oxide fuel cells *Fuel Cells* **15** 115–30
[19] Möbius H-H 1997 On the history of solid electrolyte fuel cells *J. Solid State Electrochem.* **1** 2–16
[20] Tuller H L and Barsoum M W 1985 Glass solid electrolytes: past, present and near future—the year 2004 *J. Non-Cryst. Solids* **73** 331–50
[21] Hench L L and West J K 1990 The sol–gel process *Chem. Rev.* **90** 33–72
[22] Uma T and Nogami M 2005 Influence of TiO_2 on proton conductivity in fuel cell electrolytes based on sol–gel derived P_2O_5–SiO_2 glasses *J. Non-Cryst. Solids* **351** 3325–33

[23] Carlsson J-O and Martin P M 2010 Chemical vapor deposition *Handbook of Deposition Technologies for Films and Coatings* (Amsterdam: Elsevier) pp 314–63
[24] Prakash S, Mustain W E, Park S and Kohl P A 2008 Phosphorus-doped glass proton exchange membranes for low-temperature direct methanol fuel cells *J. Power Sources* **175** 91–7
[25] Li J, Moore C W, Bhusari D, Prakash S and Kohl P A 2006 Microfabricated fuel cell with composite glass/nafion proton exchange membrane *J. Electrochem. Soc.* **153** A343
[26] Thümmler F and Thomma W 1967 The sintering process *Metall. Rev.* **12** 69–108
[27] Zeng J, He B, Lamb K, De Marco R, Shen P K and Jiang S P 2013 Anhydrous phosphoric acid functionalized sintered mesoporous silica nanocomposite proton exchange membranes for fuel cells *ACS Appl. Mater. Interfaces* **5** 11240–8
[28] Zeng J, He B, Lamb K, De Marco R, Shen P K and Jiang S P 2013 Phosphoric acid functionalized pre-sintered meso-silica for high-temperature proton exchange membrane fuel cells *Chem. Commun.* **49** 4655
[29] Nogami M, Nagao R and Wong C 1998 Proton conduction in porous silica glasses with high water content *J. Phys. Chem.* B **102** 5772–5
[30] Nogami M, Nagao R, Wong C, Kasuga T and Hayakawa T 1999 High proton conductivity in porous P2O5–SiO2 glasses *J. Phys. Chem.* B **103** 9468–72
[31] Daiko Y, Akai T, Kasuga T and Nogami M 2001 Remarkable high proton conducting P_2O_5–SiO_2 glass as a fuel cell electrolyte working at sub-zero to 120 °C *J. Ceram. Soc. Jpn.* **109** 815–7
[32] Nogami M, Suwa M and Kasuga T 2004 Proton conductivity in sol–gel-derived P_2O_5–TiO_2–SiO_2 glasses *Solid State Ion* **166** 39–43
[33] Uma T, Izuhara S and Nogami M 2006 Structural and proton conductivity study of P_2O_5–TiO_2–SiO_2 glasses *J. Eur. Ceram. Soc.* **26** 2365–72
[34] Uma T and Nogami M 2009 PMA/ZrO_2–P_2O_5–SiO_2 glass composite membranes: H_2/O_2 fuel cells *J. Memb. Sci.* **334** 123–8
[35] Uma T and Nogami M 2005 Influence of TiO_2 on proton conductivity in fuel cell electrolytes based on sol–gel derived P_2O_5–SiO_2 glasses *J. Non-Cryst. Solids* **351** 3325–33
[36] Nogami M, Matsushita H, Goto Y and Kasuga T 2000 Sol–gel-derived glass as a fuel cell electrolyte *Adv. Mater.* **12** 1370–2
[37] Suzuki S, Nozaki Y, Okumura T and Miyayama M 2006 Proton conductivity of mesoporous silica incorporated with phosphorus under high water vapor pressures up to 150 °C *J. Ceram. Soc. Jpn.* **114** 303–7
[38] Seo D, Park S, Lim B M, Cho Y S and Shul Y G 2011 Multicomponent proton conducting ceramics of SiO_2–TiO_2–ZrO_2–P_2O_5–Bi_2O_3 for an intermediate temperature fuel cell *J. Fuel Cell Sci. Technol.* **8** 011012
[39] Uma T and Nogami M 2006 On the development of proton conducting P_2O_5–ZrO_2–SiO_2 glasses for fuel cell electrolytes *Mater. Chem. Phys.* **98** 382–8
[40] Ciriminna R, Fidalgo A, Pandarus V, Béland F, Ilharco L M and Pagliaro M 2013 The sol–gel route to advanced silica-based materials and recent applications *Chem. Rev.* **113** 6592–620
[41] Elisa M, Sava B A, Volceanov A, Monteiro R c c, Alves E, Franco N, Costa Oliveira F A, Fernandes H and Ferro M C 2010 Structural and thermal characterization of SiO_2–P_2O_5 sol–gel powders upon annealing at high temperatures *J. Non-Cryst. Solids* **356** 495–501

[42] Di Z, Li H, Li M, Mao D, Chen X, Xiao M and Gu J 2012 Improved performance of fuel cell with proton-conducting glass membrane *J. Power Sources* **207** 86–90
[43] Elnahrawy A M and Ali A I 2014 Influence of reaction conditions on sol–gel process producing SiO_2 and SiO_2–P_2O_5 gel and glass *New J. Glass Ceram.* **4** 42–7
[44] Zhang L, Chae S R, Hendren Z, Park J S and Wiesner M R 2012 Recent advances in proton exchange membranes for fuel cell applications *Chem. Eng. J.* **204–205**
[45] Singhal S C and Kendall K 2003 Introduction to SOFCs *High-temperature Solid Oxide Fuel Cells: Fundamentals, Design and Applications* (Elsevier)
[46] Qing G, Kikuchi R, Takagaki A, Sugawara T and Oyama S T 2014 $CsH_5(PO_4)_2$ doped glass membranes for intermediate temperature fuel cells *J. Power Sources* **272** 1018–29
[47] Uma T and Nogami M 2007 Structural and transport properties of mixed phosphotungstic acid/phosphomolybdic acid/SiO2 glass membranes for H_2/O_2 fuel cells *Chem. Mater.* **19** 3604–10

IOP Publishing

Glass-based Materials
Advances in energy, environment and health
Sathish-Kumar Kamaraj and Arun Thirumurugan

Chapter 7

Synthesis and fabrication of quantum dots for glass windows applications

Sahaya Dennish Babu and Ananthakumar Soosaimanickam

Because of their special optical and electrical characteristics, quantum dots have attracted a lot of interest in their synthesis and manufacture for use in glass windows. An overview of the subject is given in this chapter, which covers a number of topics such as the introduction, benefits, uses, characteristics, techniques of manufacturing, and potential future developments. The rising interest in using quantum dots to improve the functionality of glass windows is covered in the introductory section. With features including high photoluminescence quantum yields, great stability, and customizable emission wavelengths, quantum dots are perfect for energy-efficient smart windows, display technologies, and photovoltaic systems. Quantum dots are used in glass windows for a variety of purposes, such as energy harvesting, colour correction, anti-reflective coatings, and better light control. Their characteristics, which include narrow emission spectra, effective light absorption, and size-dependent optical qualities, make them adaptable to a range of uses. Quantum dots in glass may be fabricated using a variety of techniques, including molecular beam epitaxy, ion exchange, sol–gel, and plasma deposition. By using these approaches, one may precisely regulate the distribution, size, and composition of quantum dots inside the glass matrix, resulting in customized optical characteristics. Quantum dots must be coated or encapsulated to prevent deterioration and to increase their compatibility with glass surfaces. Prospects for the future in this subject include investigating new encapsulating materials for long-term stability and performance enhancement, integrating with smart window technologies, and improving scalable production techniques.

7.1 Introduction

In the field of nanomaterials, quantum dots are a cutting-edge technology with special qualities and functions that may be used in a variety of ways. Quantum dots

doi:10.1088/978-0-7503-5904-7ch7

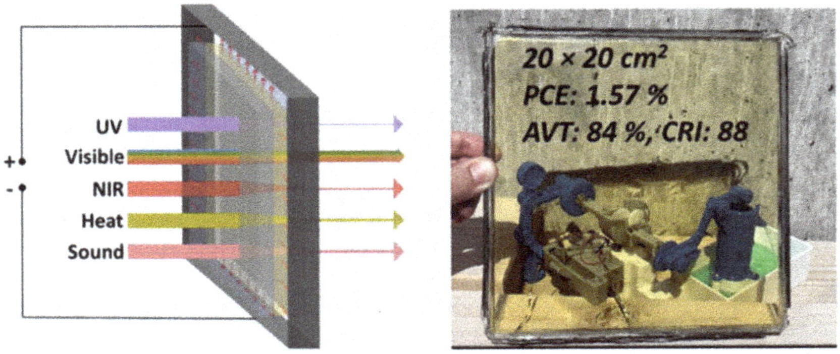

Figure 7.1. Transparent quantum dot glass and its features. Reprinted with permission from reference [3], American Chemical Society. CC BY 4.0.

open up new possibilities when they are incorporated into glass windows, turning conventional windows into cutting-edge optoelectronic devices with improved functionality and performance. Semiconductor nanoparticles with sizes in the range of a few nanometres are known as quantum dots. Quantum confinement effect has become significant at this size, resulting in distinct energy levels and optical characteristics that may be tuned. Quantum dots' optical and electrical properties are determined by their size, shape, and composition, which enables engineers to customize these nanoparticles for particular uses [1, 2]. Because they can produce light at certain wavelengths, quantum dots have impressive possibilities in the context of glass windows. Because of their tunability, windows that can amplify or filter specific colours of light may be made, which improves a building's aesthetic appeal and energy efficiency. Nowadays, quantum dots-incorporated glasses have two types: laminated glass with embedded quantum dots and edge-mounted photovoltaic (PV) cells, which are illustrated in figure 7.1. Sunlight is partly absorbed by quantum dots and is re-emitted at longer wavelengths, and travels to the edges via total internal reflection. Then, the solar cells that are mounted around the whole circumference use the fluorescence that reaches the edges and transform it into electrical power. Simultaneously, for the glazing function, a high percentage (50%) of visible light transmission must occur without haze, and the device's mechanical strength, thermal and acoustic insulation, and other features must meet the requirements of a building envelope material [2, 3].

7.2 Advantages of quantum dots in glass windows

Because quantum dots may be designed to efficiently absorb and emit light, windows that maximise natural light transmission while minimizing heat transfer can be created. This characteristic is essential for energy-efficient building designs, as windows regulate thermal comfort and lessen the need for artificial lighting and heating. Spectral selectivity in glass windows may be accomplished by varying the quantum dot size and composition. This implies that windows can provide a more pleasant and ultraviolet (UV) protected interior environment by selectively filtering out dangerous UV radiation while allowing visible light to flow through [3–5]. Dynamic light management in windows is made possible by quantum dots, which let users change the colour or tint of the glass according to their preferences or the outside environment. This function is especially helpful in smart buildings where residents may adjust the lighting to suit their needs for comfort, productivity, or aesthetics. Furthermore, transparent displays that are incorporated right into glass windows may be constructed using quantum dots as building blocks. These displays are perfect for applications like digital signage, augmented reality experiences, and smart home controls since they can display data, images, or interactive content without blocking the view.

7.3 Applications of quantum dots in glass windows

Glass windows can be made more effective in absorbing sunlight and turning it into power by adding quantum dots. By utilizing solar energy through the windows themselves, this technology—also referred to as solar windows or photovoltaic windows—can convert buildings into energy-generating structures. Quantum dots can be utilized in smart windows, which can adjust their colour or transparency in response to electrical inputs, temperature changes, and sunshine intensity, among other environmental stimuli. This makes it possible to manage light transmission dynamically, which increases energy efficiency by lowering the demand for artificial lighting, heating, and cooling [4, 5]. Glass surfaces can be coated with anti-reflective materials that contain quantum dots. These coatings are appropriate for windows in buildings, cars, and electronic displays because they minimize reflections, which minimize glare and increase sight. It is possible to construct quantum dots such that they may absorb damaging UV radiation and transform it into longer-wavelength light, protecting inhabitants from UV rays without sacrificing the transmission of visible light. This is a useful application for windows in cars, offices, and residences.

An interior space may be made more colourful and aesthetically pleasing by using quantum dots to improve the colour representation of natural light coming through windows. This is especially helpful for interior design and architectural design when colour accuracy and aesthetics are crucial. Privacy windows that alternate between transparent and opaque states can incorporate quantum dots. Based on quantum dot-activated electrochromic or photochromic effects, this technique allows for privacy management while preserving natural light transmission as needed [6–8]. When stimulated by outside sources like sunshine or artificial illumination, quantum dots implanted in glass windows can function as luminous materials. By using this idea, buildings may have energy-efficient lighting options that lessen their need on conventional light sources. It is possible to functionalize quantum dots so they can sense and react to changes in the surrounding conditions, such as temperature, humidity, or pollution levels. Real-time monitoring and management of interior environmental quality is made possible by the integration of these sensors into glass windows. Through the use of quantum dots' distinct optical and electrical characteristics, glass windows may be converted into multipurpose elements that enhance sustainability, comfort, and functionality in both cars and buildings.

7.4 Properties and characteristics of quantum dots

The size-dependent optical characteristics of quantum dots are one of its distinguishing characteristics. Discrete energy transitions result from the quantization of an electron's energy levels when a quantum dot shrinks in size. Smaller dots emit light at greater energy (shorter wavelengths) while larger dots emit light at lower energies (longer wavelengths), resulting in adjustable emission wavelengths. This characteristic is called the quantum size effect, and it is used in photovoltaics, displays, and sensors, among other uses. A semiconductor material's bandgap establishes its electrical characteristics, such as optical absorbance and conductivity. The size and makeup of the nanoparticles in quantum dots may be carefully controlled to design the bandgap. Researchers may create quantum dots with certain electrical properties, such as narrow emission spectra, high quantum efficiency, and improved charge carrier mobility, thanks to bandgap engineering [8–12]. Because of these customized characteristics, quantum dots are well suited for use in optoelectronics, where exact control over the interactions between light and matter is crucial. Quantum confinement effects are demonstrated by quantum dots, where discrete energy levels are produced by the confinement of charge carriers (holes and electrons) inside a nanoscale container. This is clearly depicted in figure 7.2. Charge carrier mobility is impeded by this confinement, which leads to increased quantum efficiency and decreased non-radiative recombination processes. Quantum dots' optical and

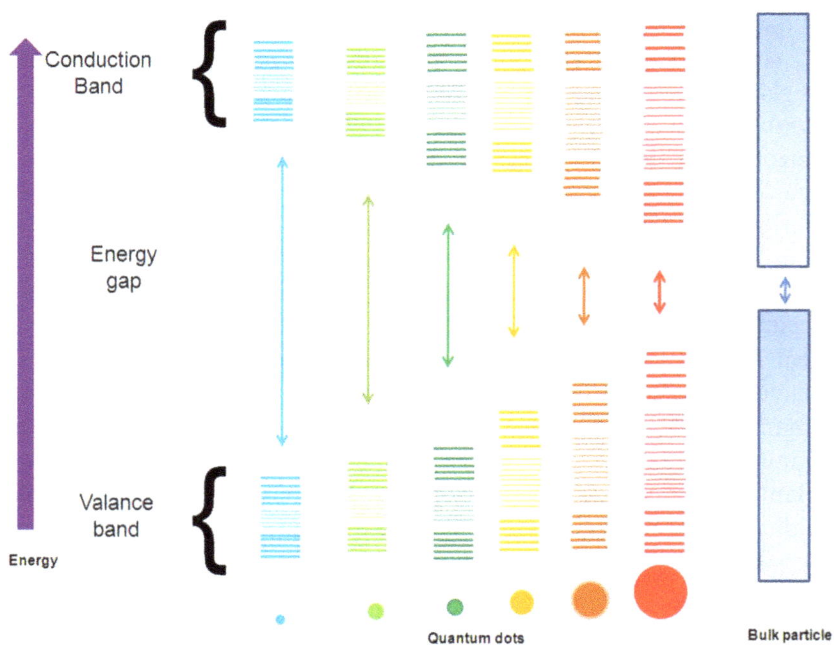

Figure 7.2. A schematic depiction of the size-tunable optical characteristics of quantum dots that shows how the emission wavelength increases as quantum dot size increases. Reprinted from [9], copyright (2019), with permission from Elsevier.

electrical characteristics are also influenced by quantum confinement phenomena, which give them a high degree of sensitivity to outside stimuli including light, electric fields, and chemical conditions. When stimulated by electricity or light, most quantum dots can produce light at particular wavelengths. The research that is now available indicates that the size and shape of quantum dots govern their electrical properties, meaning that by adjusting their size, we can regulate their emission wavelengths. Smaller quantum dots (such as those with a radius of 2–3 nm) typically emit shorter wavelengths that result in colours like violet, blue, or green. Larger quantum dots, such as those with a radius of around 5 nm, produce longer wavelengths that produce colours like orange, red, or yellow. Their highly adjustable optical capabilities due to their small size are intriguing and have prompted several scientific and industrial uses such as solar cells, LEDs, transistors, bioimaging, and diode lasers.

As we all know that widely adjustable and unique optical, electrical, chemical, and physical characteristics are present in quantum dots. They include, among other things, biology, medicine, energy harvesting, lighting, displays, cameras, sensors, communication and information technology, and sensors. These are developing in photovoltaics, sensing, and quantum information, and have been used to realize efficient lasers, displays, biotags, and solar harvesting systems accessible on the market. A quantum dot's efficiency of light emission upon stimulation is known as its quantum yield. Defect density, surface chemistry, and size are some of the factors that affect it. Smaller quantum dots often have greater quantum yields because they have fewer non-radiative energy dissipation routes. To reduce surface traps and improve radiative recombination processes, precise control over the synthesis parameters and surface functionalization is necessary to maximize the quantum yield of quantum dots [1, 6, 8]. For quantum dots to be used practically, especially in hostile settings or biological systems, their stability is essential. Quantum dot stability is increased by the use of surface passivation and encapsulation processes, which inhibit oxidation, photobleaching, and aggregation. Quantum dots are also sensitive to temperature, pH, and solvent polarity, which makes them useful for a variety of sensing and imaging applications in chemical, biological, and environmental sensing. Quantum dots give unparalleled control over light–matter interactions at the nanoscale due to their size-dependent optical characteristics, bandgap engineering capabilities, quantum confinement effects, and environmental sensitivity. Innovations in electronics, photonics, energy, healthcare, and environmental monitoring are being driven by ongoing research and development in quantum dot synthesis, surface engineering, and device integration.

7.5 Fabrication methods for quantum dots in glass

Exact methods for integrating nanoscale semiconductor particles into the glass matrix while preserving their optical and electrical characteristics are needed to fabricate quantum dots in glass. For this, a variety of techniques are used, each with their own advantages and difficulties.

7.5.1 Sol–gel method

Glass materials that are functionalized and optically transparent are produced by the sol–gel technique, which enables the uniform dispersion of quantum dots in the glass matrix. This technique provides flexibility and scalability in adjusting the concentration, size, and make-up of quantum dots in the glass. One of the most important methods for creating quantum dots in glass matrices and adding sophisticated optical properties to glass materials is the sol–gel approach. The first step in this process is to create a precursor solution that contains metal salts or metal–organic compounds, which are the building blocks needed to make quantum dots. These substances are made into sols by dissolving them in an appropriate solvent, which results in a colloidal suspension of evenly distributed nanoparticles in the liquid phase [12–14]. The sol is subjected to further alterations, such as doping with certain ions, to customize the optical and electrical characteristics of the quantum dots. Surfactants or ligands stabilize the sol and prevent particle aggregation. A number of variables, including reaction conditions, duration, and precursor concentration, may be changed to influence the size of the quantum dots. Then, the sol is gelled by means of chemical cross-linking, hydrolysis, or condensation reactions, which render the nanoparticles immobile inside a three-dimensional network assembly. After drying the gel to get rid of extra solvent, the glass containing quantum dots is consolidated and its optical properties are improved by annealing or sintering it. While partially CdS microcrystals were formed and the growth of crystallites was controlled by varying the temperature of the gas and the aeration time, as demonstrated in the work of Nogami *et al*, CdO-doped SiO_2 gel was created by dissolving $Cd(CH_3COO)_2 \cdot 2H_2O$ in a methanol solution, heat-treated at 500 °C in H_2S gas, and the gel was transformed into glass ceramics [14]. During this procedure, the size, content, crystallinity, and optical characteristics of the quantum dot glass are carefully characterized using methods including optical spectroscopy, x-ray diffraction, and transmission electron microscopy (TEM). The scalability of the sol–gel method for large-scale production combined with its precise control over quantum dot size, composition, and distribution makes it the preferred method for integrating quantum dots into glass materials for a variety of applications, from cutting-edge optical devices to energy-efficient windows. The schematic representation of processing of quantum dots and fabrication of quantum dot glass through sol–gel method is represented in figure 7.3.

7.5.2 Ion-exchange technique

One popular technology used to produce quantum dots embedded in glass matrices is ion-exchange, which provides fine control over dopant ions and makes it easier to incorporate customized optical characteristics into glass materials. The first step in this process is to create a precursor glass composition using silicon dioxide (SiO_2) as the basic ingredient. The glass is then doped with certain ions, which act as the dopants for quantum dot creation. These ions are usually transition metal ions like copper (Cu^{2+}), silver (Ag^+), or rare earth ions like europium (Eu^{3+}). The doped glass is submerged in a molten salt solution containing ions of the target quantum

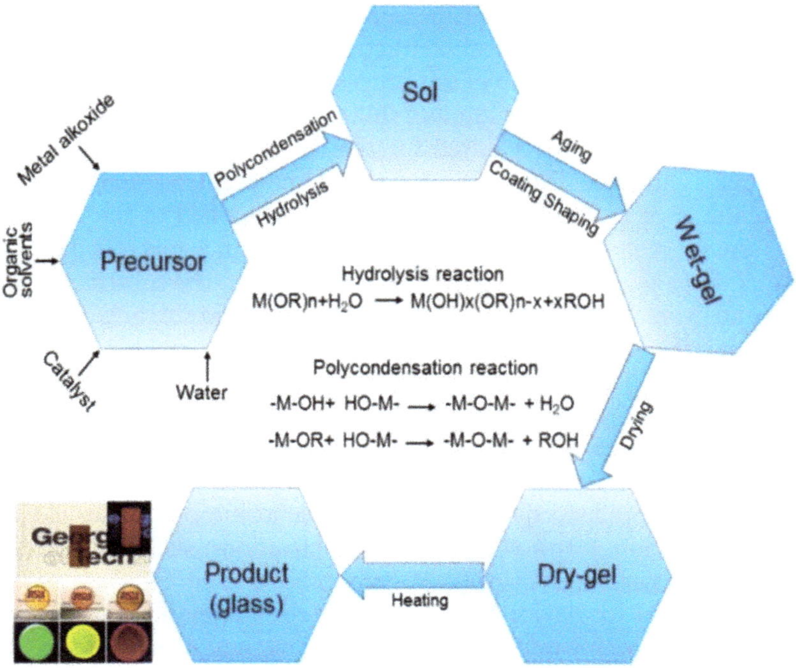

Figure 7.3. Diagram of glasses with quantum dots implanted and made using the sol–gel technique. Reprinted from [14], copyright (2020), with permission from Elsevier.

dot material as part of a controlled ion-exchange process. Quantum dots are created inside the glass structure by the ion-exchange process, which involves the interchange of dopant ions from the glass matrix with ions from the molten salt bath. The size, distribution, and concentration of quantum dots inside the glass can all be precisely controlled because of the way ions diffuse, which is dependent on their diffusion coefficients and concentration gradients [15–18]. In order to stabilize the quantum dots and enhance their optical characteristics, the glass is chilled and annealed after the ion-exchange procedure. The ion-exchange approach is a flexible way to provide sophisticated optical functions to glass materials for photonics, optoelectronics, and display applications. It may be used to customize quantum dot attributes including emission wavelength, quantum efficiency, and stability.

7.5.3 Plasma deposition

The process of plasma deposition entails producing a plasma that contains quantum dot materials and precursor gases. Quantum dots are deposited and develop on the glass substrate as a result of the plasma's interaction with the glass surface under specific circumstances. Applications needing patterned or structured quantum dot layers can benefit from the fine control that plasma deposition techniques provide over the size, density, and orientation of quantum dots on glass. A highly developed method for creating quantum dots on glass substrates is plasma deposition, which allows for exact control over the parameters of the deposition process and allows for the incorporation of specific optical characteristics of the quantum dots into glass materials [8, 15–18]. The first step in this process is to prepare a target material, such as semiconductor materials like cadmium sulfide (CdS) or cadmium selenide (CdSe), that contains the ingredients required for the creation of quantum dots. Next, a plasma source is used to evaporate the target material, producing a stream of evaporated atoms or molecules. This vapour flux is applied to the glass substrate, which has usually been cleaned beforehand and is kept in a controlled environment. This permits the deposition of quantum dot material on the substrate surface. The nucleation and development of nanocrystals with regulated size and composition are encouraged by the energetically favourable conditions created by the plasma environment, which leads to the production of quantum dots. The deposition temperature, duration, gas composition, and plasma power may all be adjusted to precisely regulate the size, density, and distribution of quantum dots within the glass substrate [18–20]. The optical characteristics and stability of the glass containing quantum dots can be enhanced by applying post-deposition processes like annealing or surface modification. For a wide range of applications, such as optical coatings, sensors, photovoltaics, and displays, where precise control over quantum dot properties is essential for device performance and functionality, plasma deposition is the preferred method for integrating quantum dots into glass materials due to its versatility and scalability.

7.5.4 Molecular beam epitaxy

A high-precision method for depositing tiny layers of quantum dot materials onto a substrate, such as glass is called molecular beam epitaxy (MBE). In MBE, controlled growth procedures cause molecular beams of elemental or compound semiconductors to create epitaxial layers of quantum dots on the glass surface. The size, composition, and form of quantum dots may be precisely controlled at the atomic level with MBE, resulting in glass quantum dot structures that are well-defined and of superior quality. An ultra-high vacuum environment is first created as part of the MBE process in order to reduce contaminants and provide exact control over material deposition [8, 14, 19]. A substrate is manufactured and put within the vacuum chamber. This substrate is often comprised of a semiconductor material, such as silicon (Si) or gallium arsenide (GaAs). After being heated to create a molecular stream of atoms or molecules, the required quantum dot material, typically semiconductor compounds like indium arsenide (InAs) or gallium nitride (GaN), is extracted. Atomic layers are deposited on the substrate surface when these molecular beams are directed towards it with regulated energy and flow. When it comes to quantum dots in glasses, the MBE procedure entails covering a glass substrate with thin layers of semiconductor materials that include the components of a quantum dot. By carefully regulating the flow of atoms or molecules in the molecular beams, this deposition is made possible, enabling the development of nanoscale quantum dot patterns on the glass surface. Bao *et al* prepared integrated

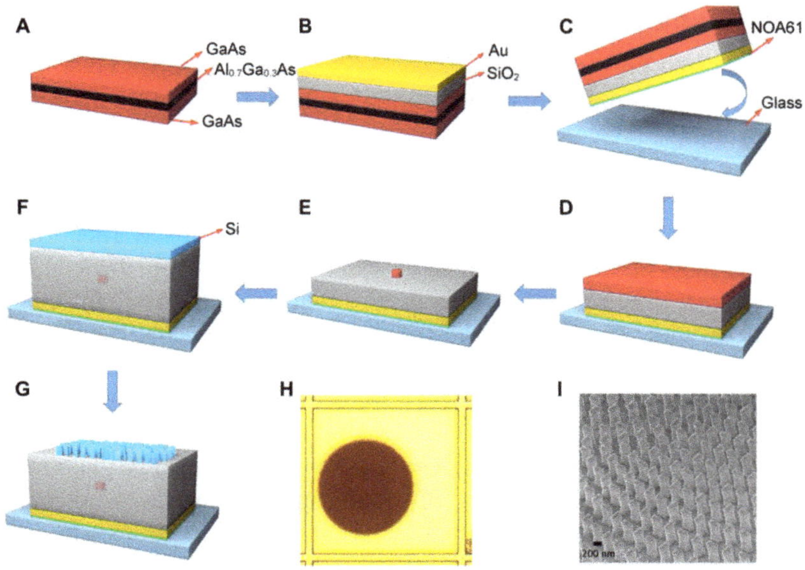

Figure 7.4. Fabrication process of integrated quantum dots on glass substrates using MBE method. Reprinted from [21] with permission from AAAS.

quantum dots on glass substrates for the possible applications in spin-splitting bifocal metalens, which is depicted in figure 7.4. Firstly, a single layer of InAs quantum dots is generated using MBE to create a 160-nm-thick GaAs membrane, after which a 1-μm-thick $Al_{0.7}Ga_{0.3}As$ sacrificial layer is formed on a GaAs substrate. Subsequently, a 1-μm thick SiO_2 layer was applied to the wafer's surface, and then a 1-nm-thick Ti adhesion layer and a 100-nm-thick reflective Au layer were evaporated using an e-beam. The wafer was attached to a clear glass by spinning and UV-curing adhesive NOA61 (Norland Products Inc.). Next, using a selective wet-etching technique, the GaAs substrate and $Al_{0.7}Ga_{0.3}As$ sacrificial layer are eliminated, leaving the GaAs membrane at the top [21].

In order to guarantee the best possible adhesion, crystallinity, and alignment of the quantum dots inside the glass matrix, the substrate temperature is meticulously regulated during the deposition process. The ability of MBE to construct quantum dots with precise size, shape, and composition is one of its main benefits for quantum dot synthesis. Through the manipulation of deposition parameters, such as substrate temperature, beam flux, and growth time, scientists may precisely regulate the size of quantum dots in the nanometer scale range, resulting in optical and electrical characteristics that can be tuned. MBE is a favoured technique for research and development in quantum dot-based devices, such as lasers, light-emitting diodes (LEDs), solar cells, and quantum dot displays, due to its remarkable control over quantum dot growth [22]. MBE is a potent tool for developing semiconductor optoelectronics and nanotechnology because of its capacity to produce very homogeneous, flawless, and precisely tunable quantum dot layers.

7.6 Coating and encapsulation

Coating or encasing pre-synthesized quantum dots in a protective layer that may be included into the glass matrix is an additional method for creating quantum dots in glass. This technique offers stability and glass substrate compatibility while assisting in maintaining the optical and electrical characteristics of quantum dots. Polymer coatings, silica shells, and organic–inorganic hybrid materials are examples of coating and encapsulating approaches that provide improved functionality and durability for glass materials boosted by quantum dots [16]. The main goal of coating and encapsulation is to create a barrier that protects the quantum dots from outside elements that might eventually weaken their stability and performance, such as moisture, oxygen, and chemicals. The quantum dots' sensitivity to external factors is reduced by encapsulation, which enhances their lifetime and dependability in a variety of applications.

Coating and encapsulation are frequently used in the context of glass materials to improve the optical characteristics of quantum dots, such as their quantum efficiency, emission wavelength, and photostability. Quantum dot surface chemistry can be altered by coatings, resulting in enhanced dispersibility, matrix compatibility, and decreased aggregation tendencies. This guarantees uniform distribution and excellent performance inside the glass matrix, which is especially crucial when integrating quantum dots onto glass substrates. Depending on the required qualities and application requirements, several procedures are applied for coating and encapsulating quantum dots in glasses [18–20]. For instance, to increase the durability and biocompatibility of quantum dots for use in biological and medicinal applications, polymer coatings like polyethylene glycol (PEG), polyvinyl alcohol (PVA), or silica-based coatings can be added. These coatings not only shield the quantum dots from deterioration but also make it easier for them to be incorporated into particular host materials like glasses without substantially changing their optical characteristics. For quantum dot encapsulation in glasses, inorganic encapsulation techniques like chemical vapour deposition (CVD), atomic layer deposition (ALD), or shell formation utilising sol–gel procedures are frequently used in addition to polymer coatings. By using these techniques, uniform and conformal shells that offer mechanical support, chemical resistance, and improved optical qualities may be deposited around the quantum dots. For instance, sol–gel methods may be used to create a silica shell around quantum dots, which offers superior encapsulation, optical transparency, and compatibility with glass substrates [19–22]. In addition, it is possible to modify the functionalization of coatings and encapsulating layers in order to provide quantum dots certain characteristics like improved luminescence, water solubility, or biocompatibility. For uses in biosensing, imaging, drug delivery, and optoelectronics, surface modifications using ligands, biomolecules, or targeted moieties can also enable particular interactions and functions.

7.7 Future prospects and developments

The potential for quantum dot-glass window technology to revolutionize the energy efficiency, usability, and aesthetics of buildings and automobiles is enormous.

Thanks to its numerous advantages and breakthroughs in a variety of fields, quantum dot-glass windows have the potential to become essential parts of intelligent, sustainable infrastructure. Energy harvesting and efficiency are major areas of interest for future advances in quantum dot-glass window technology. Quantum dot-glass windows have the potential to turn buildings into self-sufficient energy producers by integrating quantum dots that can absorb solar radiation and turn it into electricity. This technique, sometimes referred to as solar windows or photovoltaic windows, helps to significantly reduce carbon emissions and the impact on the environment in addition to lowering reliance on conventional energy sources. Furthermore, enhanced optical characteristics and performance of quantum dot-glass windows are anticipated as a result of developments in quantum dot engineering and manufacturing techniques. Accurate manipulation of quantum dot size, composition, and distribution will allow windows to be customized with desired optical properties, such as improved colour rendering, light manipulation, and transparency. This will create new opportunities for daylighting techniques in buildings, interior design, and architectural design.

The incorporation of smart features into quantum dot-glass windows is another innovative field. Subsequent advancements might involve the integration of sensors, actuators, and nanoscale electronics straight into the glass matrix, enabling real-time observation, management, and adjustment of window characteristics in reaction to external circumstances. Furthermore, enhanced optical characteristics and performance of quantum dot-glass windows are anticipated as a result of developments in quantum dot engineering and manufacturing techniques. Accurate manipulation of quantum dot size, composition, and distribution will allow windows to be customized with desired optical properties, such as improved colour rendering, light manipulation, and transparency. This will create new opportunities for daylighting techniques in buildings, interior design, and architectural design. The incorporation of smart features into quantum dot-glass windows is another innovative field. Subsequent advancements might involve the integration of sensors, actuators, and nanoscale electronics straight into the glass matrix, enabling real-time observation, management, and adjustment of window characteristics in reaction to external circumstances. All things considered, quantum dot-glass window technology has enormous potential to transform the built environment in the future by providing clever, sustainable, and aesthetically beautiful solutions that make the world a greener and smarter place. Realizing these futuristic ideas and realizing the full benefits of quantum dot-glass window technology for society will need sustained research, investment, and cooperation.

7.8 Conclusion

In summary, investigating quantum dots as a means of improving glass windows offers a viable path towards practical and energy-efficient architectural solutions. This chapter explores the introduction, benefits, uses, characteristics, processes of production, coating, and possibilities for the future of quantum dots in glass. Their distinct optical and electrical characteristics, including size-dependent optical quality, adjustable emission wavelengths, and high photoluminescence quantum yields, make them perfect for a range of applications, including anti-reflective coatings, light control, and energy harvesting. Molecular beam epitaxy, ion exchange, sol–gel, and plasma deposition are just a few of the varied production methods that provide fine control over the distribution and properties of quantum dots inside the glass matrix. Their compatibility with glass surfaces and long-term stability are guaranteed by encapsulation and coating techniques. Prospective paths comprise investigating novel encapsulating materials, incorporating smart window technologies, and optimizing production methods for expandable uses. All things considered, quantum dots have great potential to transform the sustainability and usability of glass windows in the technical and architectural spheres.

Acknowledgments

One of the authors SDB sincerely thanks the management, Chettinad College of Engineering and Technology, Karur for their extended support during this work.

References

[1] Xia M, Luo J, Chen C, Liu H and Tang J 2019 Semiconductor quantum dots-embedded inorganic glasses: fabrication, luminescent properties, and potential applications *Adv. Opt. Mater.* **7** 1900851

[2] AbouElhamd A R, Al-Sallal K A and Hassan A 2019 Review of core/shell quantum dots technology integrated into building's glazing *Energies* **12** 1058

[3] Huang J, Zhou J, Jungstedt E, Samanta A, Linnros J, Berglund L A and Sychugov I 2022 Large-area transparent 'quantum dot glass' for building-integrated photovoltaics *ACS Photonics* **9** 2499–509

[4] Makarov N S, Korus D, Freppon D, Ramasamy K, Houck D W, Velarde A, Parameswar A, Bergren M R and McDaniel H 2022 Minimizing scaling losses in high-performance quantum dot luminescent solar concentrators for large-area solar windows *ACS Appl. Mater. Interfaces* **14** 29679–89

[5] Bera D, Qian L, Tseng T-K and Holloway P H 2010 Quantum dots and their multimodal applications: a review *Materials* **3** 2260–345

[6] Ma Y, Zhang Y and Yu William W 2019 Near infrared emitting quantum dots: synthesis, luminescence properties and applications *J. Mater. Chem.* C **7** 13662–79

[7] Nozik A J, Beard M C, Luther J M, Law M, Ellingson R J and Johnson J C 2010 Semiconductor quantum dots and quantum dot arrays and applications of multiple exciton generation to third-generation photovoltaic solar cells *Chem. Rev.* **110** 6873–90

[8] Xie B, Hu R and Luo X 2016 Quantum dots-converted light-emitting diodes packaging for lighting and display: status and perspectives *J. Electron. Packag.* **138** 020803

[9] Reshma V G and Mohanan P V 2019 Quantum dots: applications and safety consequences *J. Lumin.* **205** 287–98

[10] Cao H, Ma J, Huang L, Qin H, Meng R, Li Y and Peng X 2016 Design and synthesis of antiblinking and antibleaching quantum dots in multiple colors via wave function confinement *JACS* **138** 15727–35

[11] Zhong J, Zhao H, Zhang C, Ma X, Pei L, Liang X and Xiang W 2014 Sol–gel synthesis and optical properties of $CuGaS_2$ quantum dots embedded in sodium borosilicate glass *J. Alloys Compd.* **610** 392–8
[12] Selvan S T, Bullen C, Ashokkumar M and Mulvaney P 2001 Synthesis of tunable, highly luminescent QD-glasses through sol-gel processing *Adv. Mater.* **13** 985–8
[13] Nogami M, Nagasaka K and Takata M 1990 CdS microcrystal-doped silica glass prepared by the sol-gel process *J. Non-Cryst. Solids* **122** 101–6
[14] Xue J, Wang X, Jeong J H and Yan X 2020 Fabrication, photoluminescence and applications of quantum dots embedded glass ceramics *Chem. Eng. J.* **383** 123082
[15] Li J, Dong H, Zhang S, Ma Y, Wang J and Zhang L 2016 Colloidal quantum-dot-based silica gel glass: two-photon absorption, emission, and quenching mechanism *Nanoscale* **8** 16440–8
[16] Xu K and Heo J 2012 Effect of silver ion-exchange on the precipitation of lead sulfide quantum dots in glasses *J. Am. Ceram. Soc.* **95** 2880–4
[17] Chen D, Yuan S, Chen J, Zhong J and Xu X 2018 Robust $CsPbX_3$ (X = Cl, Br, and I) perovskite quantum dot embedded glasses: nanocrystallization, improved stability and visible full-spectral tunable emissions *J. Mater. Chem.* C **6** 12864–70
[18] Lin J, Lu Y, Li X, Huang F, Yang C, Liu M, Jiang N and Chen D 2021 Perovskite quantum dots glasses based backlit displays *ACS Energy Lett.* **6** 519–28
[19] Pchelyakov O P, Nikiforov A I, Olshanetsky B Z, Teys S A, Yakimov A I and Chikichev S I 2008 MBE growth of ultra small coherent Ge quantum dots in silicon for applications in nanoelectronics *J. Phys. Chem. Solids* **69** 669–72
[20] Yoffe A D 2001 Semiconductor quantum dots and related systems: electronic, optical, luminescence and related properties of low dimensional systems *Adv. Phys.* **50** 1–208
[21] Bao Y, Lin Q, Su R, Zhou Z-K, Song J, Li J and Wang X-H 2020 On-demand spin-state manipulation of single-photon emission from quantum dot integrated with metasurface *Sci. Adv.* **6** eaba8761
[22] Hjiri M, Hassen F and Maareference H 2000 Optical characterisation of self organized InAs/GaAs quantum dots grown by MBE *Mater. Sci. Eng.* B **74** 253–8

IOP Publishing

Glass-based Materials
Advances in energy, environment and health
Sathish-Kumar Kamaraj and Arun Thirumurugan

Chapter 8

Technological advancements of photocatalysis on glass substrates

Sahaya Dennish Babu George and Swetha Madamala

Using glass substrates for photocatalysis has shown promise in resolving environmental issues and improving sustainability in a variety of sectors. The basics of photocatalysis are examined in this chapter with a focus on its applications in environmental remediation, antimicrobial surfaces, and air and water purification. Glass substrates provide stability, endurance, and increased photocatalytic activity, making them the ideal base for photocatalyst coatings. A great deal of progress has been made in material science and engineering, from laboratory-scale research to scalable manufacturing methods, as evidenced by milestones in the creation of photocatalyst coatings on glass substrates. Photocatalysts may be precisely and uniformly applied onto glass substrates using a variety of deposition processes, such as chemical vapor deposition (CVD) and sol–gel procedures. This allows for customization and adaptability across a broad range of applications. Photocatalyst-glass composites, which provide solutions for water purification, indoor air quality enhancement, and self-cleaning surfaces, represent a convergence of technology and sustainability. Durability, scaling up, and commercialization are only a few of the difficulties that make this subject more in need of research and innovation. Energy-efficient buildings and sustainable development projects are made possible by the integration of smart glass technology, which creates new opportunities for intelligent surfaces with adaptive features. The chapter on photocatalyst technical breakthroughs on glass substrates explains how this technology may drastically improve the quality of air and create healthier, cleaner surroundings. It emphasises how crucial innovation and teamwork are to advancing the field and maximizing the potential of photocatalyst coatings combined with glass substrates.

8.1 Introduction

In modern studies, photocatalysis is a key mechanism that promises revolutionary effects in a number of fields, including energy generation and environmental

doi:10.1088/978-0-7503-5904-7ch8

cleanup. Photocatalysis is fundamentally the result of an intricate interaction between light, semiconductors, and chemical processes. At its core, it is the process of using light energy to promote chemical reactions on the surface of a catalyst, usually a semiconductor. Thanks to its resilience and photoactivity, titanium dioxide (TiO_2) is one of the most researched photocatalysts [1–3]. When light's photons hit the photocatalyst's surface, they excite electrons in the valence band to move into the conduction band, starting the process of forming electron–hole pairs. These charged species facilitate the breakdown of organic contaminants or the splitting of water molecules to produce hydrogen by acting as active sites for redox reactions with nearby species. Because of photocatalysis's extraordinary adaptability, it may be used in a wide range of settings, such as the treatment of wastewater, air purification, and solar energy conversion. Nevertheless, issues like low quantum efficiency and catalyst stability continue to exist, which motivates further research projects meant to clarify the complexities of photocatalytic operations and improve material design approaches. Photocatalysis highlights the tremendous synergy between light and matter in promoting chemical transformations with broad implications for a more sustainable and greener future, in addition to unlocking the possibilities for sustainable innovations. Due to its potential uses in energy production, water purification, environmental remediation, and even organic chemical synthesis, photocatalysis has attracted a great deal of attention [4, 5]. The fundamental idea behind photocatalysis is using photon energy to start chemical reactions. It promises to provide environmentally friendly and long-lasting answers to a number of human problems, such as pollution and energy scarcity. Understanding the fundamentals of surface chemistry, semiconductor physics, and reaction kinetics is necessary to comprehend the mechanics of photocatalysis. The photocatalyst, a semiconductor substance with the ability to absorb light energy and start chemical reactions, is the central component of photocatalysis. Electrons in the material are excited and moved from the valence band to the conduction band when photocatalyst surface is exposed to light with enough energy. Electrons (e^-) and positively charged holes (h^+), two very reactive species, are produced by this process as electron–hole pairs [6, 7].

The essential intermediaries that propel photocatalytic processes are these pairs of electrons and holes. While holes in the valence band have oxidizing characteristics and can aid in oxidation events, electrons in the conduction band have reducing potential and can take part in reduction reactions. A sequence of redox reactions that the photocatalyst catalyses are part of the photocatalytic process. One of the most well-known applications of photocatalysis is the removal of organic contaminants from air and water. Pesticides, dyes, and volatile organic compounds (VOCs) are examples of organic pollutants that provide serious environmental risks. One effective and sustainable way to break them down is by photocatalysis [8]. Organic contaminants adsorb onto the surface of the semiconductor material when a photocatalyst and light are present. Following light absorption, the excited electrons and holes produced interact with the pollutants that have been adsorbed, starting a series of oxidation and reduction processes. In the end, this causes organic substances to mineralize and produce innocuous byproducts like carbon dioxide and water. The production of renewable energy by the splitting of water molecules is another significant use of photocatalysis. This method, called photocatalytic water splitting, uses sunlight as an energy source to help the photocatalyst convert water into hydrogen and oxygen gases. With its high energy content and pure nature, hydrogen has great promise as a long-term replacement for fossil fuels. Although photocatalysis has a lot of potential, there are a few issues that need to be resolved before it can be fully utilized. The comparatively poor efficiency of photocatalytic processes is one of the main drawbacks. A few examples of the variables that affect photocatalysis efficiency include the semiconductor material's bandgap, surface area, crystal structure, and light absorption capabilities. To increase the effectiveness of photocatalytic processes, new photocatalysts with improved light absorption and charge separation characteristics must be developed [9]. To improve the performance of photocatalytic materials, researchers are investigating a number of tactics, including surface modification, heterostructure creation, and metal ion doping. In addition, difficulties arise from the photocatalytic processes' selectivity and specificity. Unwanted side reactions can happen in complicated reaction settings and cause the photocatalyst to deactivate or create byproducts. Reaction conditions must be optimized, and reaction mechanisms comprehended in order to achieve high selectivity in photocatalytic processes [10]. Concerns still exist regarding the photocatalytic technology's affordability and scalability. Although photocatalysis has been shown to be effective in laboratory settings for energy production and environmental cleanup, overcoming technical and financial obstacles is necessary to translate these results into larger scale practical applications.

8.2 Importance of photocatalysis in various industries

8.2.1 Environmental remediation

With its capacity to use light energy to propel chemical processes, photocatalysis has become a vital technology with broad applications in a variety of sectors. Its significance goes beyond simple chemical reactions; it affects everything from healthcare and energy generation to environmental cleanup. Environmental cleanup is one of the most common uses of photocatalysis. As worries about pollution and the depletion of natural resources grow, the necessity for efficient remediation technology has increased. A sustainable solution is provided by photocatalysis, which uses sunshine to break down dangerous chemicals found in soil, water, and the air. Nitrogen oxides (NO_x), VOCs, and other dangerous airborne pollutants can be catalysed to break down into harmless byproducts like carbon dioxide and water by photocatalytic materials like titanium dioxide (TiO_2) [4–8]. Urban areas can reduce air pollution and enhance air quality by applying photocatalytic coatings on surfaces that are exposed to sunlight, such as sidewalks and building facades. Similar to this, photocatalysis is essential for treating wastewater, as traditional techniques frequently fail to handle new contaminants and stubborn organic pollutants. Under UV or visible-light irradiation, organic chemicals, pathogens, and pharmaceutical residues found in wastewater streams are broken down by semiconductor photocatalysts in photocatalytic reactors. This technique reduces the environmental impact of wastewater discharge and protects water resources by providing an affordable and sustainable alternative to conventional treatment methods.

8.2.2 Healthcare and biomedical applications

Photocatalysis has great potential in biological research and healthcare to treat microbial infections and provide novel therapeutic therapies. Because photocatalytic materials are naturally antibacterial, they may effectively combat a wide range of pathogens, such as bacteria, viruses, and fungus. By continually destroying bacteria upon exposure to light, photocatalytic coatings incorporated into hospital surfaces, medical devices, and personal protective equipment might help decrease the incidence of healthcare-associated infections (HAIs). By reducing the need for chemical disinfectants, this proactive strategy not only improves patient safety but also lowers the risk of environmental pollution and antimicrobial resistance [11]. Moreover, photocatalysis is used in the creation of photodynamic therapy (PDT), a non-invasive cancer and other illness treatment method. In photodynamic therapy (PDT), photosensitizing chemicals are injected into target tissues and then selectively accumulated there. This is followed by the application of light at particular wavelengths, which causes cytotoxic responses and kills cancerous cells. Semiconductor-based photocatalysts with adjustable optical characteristics and biocompatibility, such as quantum dots and nanoscale metal oxides, have been investigated as possible PDT photosensitizers [12]. Researchers want to improve PDT's selectivity and efficacy while reducing harmful effects on healthy tissues by using the principles of photocatalysis.

8.2.3 Energy conversion and storage

In the search for renewable energy sources and the move towards a carbon-neutral economy, photocatalysis is essential. Photocatalytic systems provide a sustainable method for energy conversion and storage by using solar radiation to drive chemical processes. This has significant implications for the production of fuel, power, and environmental sustainability. Photocatalytic water splitting is a viable method for producing hydrogen, a clean and multipurpose energy source, in the field of solar fuel synthesis [3]. When exposed to natural or artificial light, semiconductor photocatalysts can split water molecules into hydrogen and oxygen gases. This process produces renewable hydrogen fuel that can be used in fuel cells, vehicles, and industrial processes, among other applications. Even if there are still many obstacles to overcome in order to maximize the stability and efficiency of photocatalytic water splitting systems, research is still being done to promote innovation and the commercialization of this technology. Additionally, through artificial photosynthesis, photocatalysis shows promise for the utilization of carbon dioxide (CO_2) and its conversion into chemicals and fuels with additional value [7]. Photocatalytic reactors can help produce carbon-neutral fuels like methanol, methane, and hydrocarbons by combining the reduction of CO_2 with renewable energy sources like solar or wind power. This reduces greenhouse gas emissions and aids in the fight against climate change.

Photocatalysis is a flexible platform used in materials science and nanotechnology to synthesize, functionalize, and characterize new materials with specific features and uses. Precision control over chemical transformations and material characteristics is made possible by the unique optical, electrical, and catalytic features of semiconductor photocatalysts, which are created at the nanoscale. The remarkable photocatalytic activity, surface reactivity, and programmable shape of nanomaterials like quantum dots, nanowires, and mesoporous metal oxides have attracted a lot of interest [5, 9]. Compared to their bulk counterparts, these nanostructured photocatalysts have improved light absorption, charge separation, and catalytic efficiency, which makes them ideal for a variety of uses, such as environmental remediation, photovoltaics, sensors, and catalysis. Moreover, photocatalysis is an effective technique for creating thin films and functional coatings with specific surface qualities, such as anti-corrosive, anti-fouling, and self-cleaning qualities. Researchers can add new functions to a variety of surfaces, such as glass, ceramics, metals, and polymers, by depositing photocatalytic coatings onto substrates using methods like CVD, atomic layer deposition (ALD), and sol–gel processing. This will improve the surfaces' performance and durability in a range of applications. It is impossible to overestimate the significance of photocatalysis in a variety of sectors as it fosters innovation, solves global issues, and opens doors for a more sustainable future. Photocatalytic technologies offer flexible solutions for a wide range of applications, with significant consequences for human health, environmental stewardship, and economic success. These applications span from environmental remediation and healthcare to energy conversion and materials research [11, 12]. The incorporation of photocatalysis into daily applications and industrial processes, as research and technology develop, has the potential to open up new avenues and create a more sustainable and optimistic environment for future.

8.3 Glass substrates in photocatalytic applications

8.3.1 Properties of glass substrates

Glass substrates are essential in many domains, from optics to electronics, because of their special qualities and adaptability. These substrates, which are usually composed of materials based on silica, have a wide range of mechanical, optical, and physical properties that make them essential for use in a variety of applications. Figure 8.1 represents the essential properties of glass substrates which suitable for photocatalytic applications.

8.3.1.1 Transparency and optical clarity

Transparency and optical clarity are two of glass substrates' most distinctive characteristics. Glass is the perfect material for situations where optical transparency is essential since it lets light travel through it with little absorption or scattering [11]. In the manufacturing of optical lenses, photovoltaic devices, and display panels, this characteristic is especially helpful. Glass substrates allow light to pass through with little distortion, guaranteeing excellent optical performance and imaging.

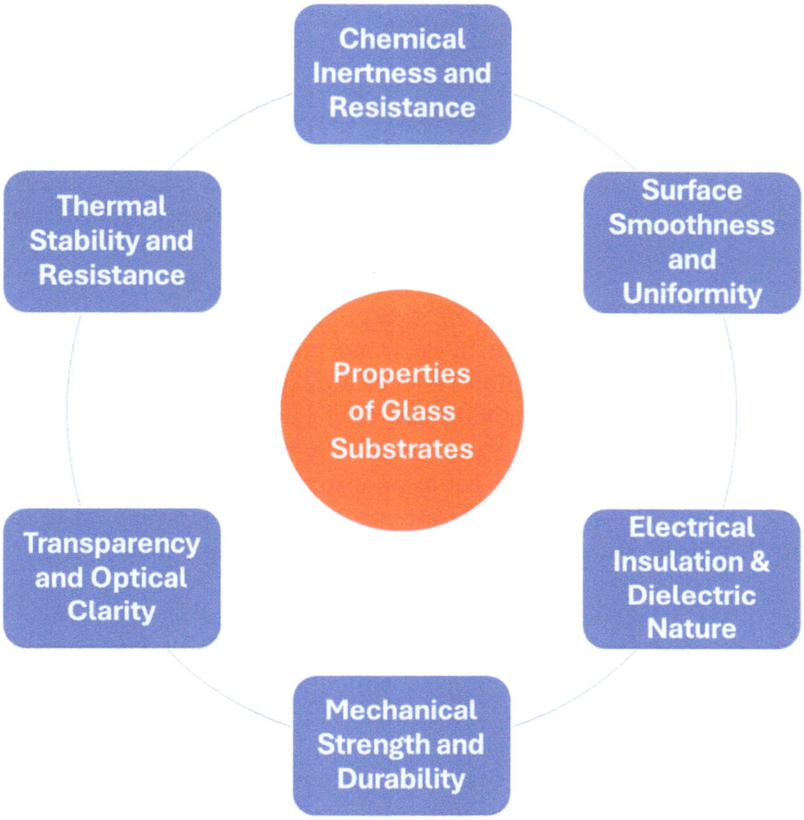

Figure 8.1. List of properties of glass substrates.

8.3.1.2 Thermal stability and resistance

Glass substrates are very resistant to temperature changes and have good thermal stability. They are appropriate for use in high-temperature situations because they can tolerate high temperatures without deforming or sustaining structural damage. This feature is especially crucial in industries like semiconductor production, where accurate temperature control is necessary to ensure the dependability and performance of the device [12–14]. For the deposition and patterning of thin films used in optoelectronics and microelectronics, glass substrates offer a sturdy foundation.

8.3.1.3 Chemical inertness and resistance

Glass substrates are very resistant to corrosion from acids, bases, and other reactive chemicals, demonstrating their great chemical inertness. They can be used in difficult chemical settings where other materials may break down or react with the surrounding medium because of this feature [12–14]. Glass substrates are frequently used in chemical processing facilities, laboratory apparatus, and medicinal devices where chemical resistance is essential. The lifetime and integrity of components exposed to corrosive chemicals are guaranteed by the inert nature of glass.

8.3.1.4 Surface smoothness and uniformity

Glass substrates are appropriate for applications requiring precision patterning and thin layer deposition because of their uniformly smooth surface texture. Glass substrates' homogeneous coating deposition and adherence are made possible by their smoothness and flatness, which is crucial for the construction of optical and electrical devices [13]. Glass substrates provide higher surface quality, which reduces flaws and irregularities and improves the overall dependability and performance of produced components.

8.3.1.5 Electrical insulation and dielectric properties

Glass substrates are perfect for use in electronic and semiconductor devices because of their superior electrical insulation and dielectric qualities. Because glass materials have a limited electrical conductivity, electrical components are kept isolated and current leaks are avoided. This feature is especially helpful for making printed circuit boards (PCBs), as glass substrates act as an insulating barrier between components and conductive lines [14]. Glass substrates provide dependable electrical separation while allowing electronic circuit integration and downsizing.

8.3.1.6 Mechanical strength and durability

Glass substrates demonstrate high mechanical strength and durability, enabling them to withstand mechanical stress and impact without fracture or failure. Glass materials exhibit a combination of hardness and toughness, making them resilient to scratches, abrasion, and physical damage. This property is essential for applications such as automotive glazing, architectural windows, and protective covers where resistance to mechanical forces is paramount [15]. Glass substrates provide structural integrity and safety in various environments, ensuring long-term performance and reliability.

Depending on the structure and content of the glass material, glass substrates display different optical characteristics, such as dispersion, refractive index, and light transmission characteristics. These optical characteristics can be modified to satisfy the needs of certain applications, including optical waveguides, lenses, and filters. Glass substrates provide exact control over the manipulation and propagation of light, which makes it possible to create cutting-edge optical systems and devices for sensing, imaging, and communications [16]. Glass substrates are an incredibly adaptable and useful material platform with a variety of qualities appropriate for a wide range of uses. They are perfect for usage in the electronics, optics, aerospace, automotive, and biomedical sectors due to their transparency, thermal stability, chemical resistance, surface smoothness, electrical insulation, mechanical strength, and optical qualities [14–16]. The need for glass substrates with customized qualities and improved performance will only increase as technology develops and new applications appear, spurring creativity and progress in materials science and engineering.

8.3.2 Role of glass substrates in photocatalysis

The photocatalyst, a semiconductor substance with the ability to absorb light energy and start chemical reactions, is the central component of photocatalysis. Glass substrates are an essential part of photocatalytic systems because they support photocatalytic materials and enable effective light absorption. Recent years have seen an increase in the need for efficient pollution control techniques and sustainable energy sources, which has fueled photocatalysis research and development. Photocatalytic systems are a sustainable way to solve environmental problems including air and water pollution, greenhouse gas emissions, and energy scarcity by using the power of sunshine. Glass substrates can be combined with photocatalytic materials to create effective, scalable photocatalytic systems that can be used for a variety of purposes [8, 13]. In the realm of photocatalysis, glass substrates are essential because they provide the support and performance enhancement for photocatalytic materials. Photocatalysis is the process of using light energy to start chemical reactions. It has a lot of potential uses in energy generation, environmental cleanup, and other industrial operations. The effective use of light energy and the development of scalable and sustainable photocatalytic systems are made possible by the combination of photocatalytic materials with glass substrates. Glass substrates offer a robust and inert surface on which photocatalytic compounds may be deposited and rendered immobile. Solid support is necessary for photocatalysts, which are usually semiconductor materials like zinc oxide (ZnO) or titanium dioxide (TiO_2), to preserve their structural integrity and enable effective light absorption [6]. Because glass substrates are transparent to ultraviolet (UV) and visible light, incident light can pass through and reach the photocatalytic layer underneath. By igniting the photocatalyst and encouraging the creation of electron–hole pairs, transparency is essential for starting photocatalytic processes. Figure 8.2 provides the list of roles of glass substrates in photocatalysis applications.

Glass substrates ensure optimal light transmission and utilization by the photocatalytic material due to their negligible light absorption and scattering. Glass substrates' optical clarity increases photocatalytic processes' efficacy and efficiency by facilitating the breakdown of pollutants and the production of reactive oxygen species when exposed to light [10, 13]. The surface homogeneity and smoothness of glass substrates are two characteristics that might affect the supported materials' photocatalytic activity. The photocatalyst and substrate will make the best possible contact when the glass substrate is smooth and has been properly prepared. At the interface between the photocatalyst and substrate, this homogeneity reduces surface imperfections and facilitates effective charge transfer and reaction kinetics. The system's overall stability and photocatalytic activity are increased by the synergistic interaction of the photocatalytic material and glass substrate [11–14]. Glass substrates are a good choice for long-term photocatalysis applications because of their exceptional durability and resistance to chemical degradation. Glass substrates do not break down or react with photocatalytic compounds or the environment, in contrast to organic substrates. For longer durations, they retain their optical characteristics and structural integrity, guaranteeing the dependability and durability of photocatalytic systems. Glass substrates offer photocatalytic materials a stable and inert support platform that allows for repeated usage and continuous functioning in a variety of industrial and environmental contexts. Glass substrates may be used in a wide range of photocatalytic designs and applications because of

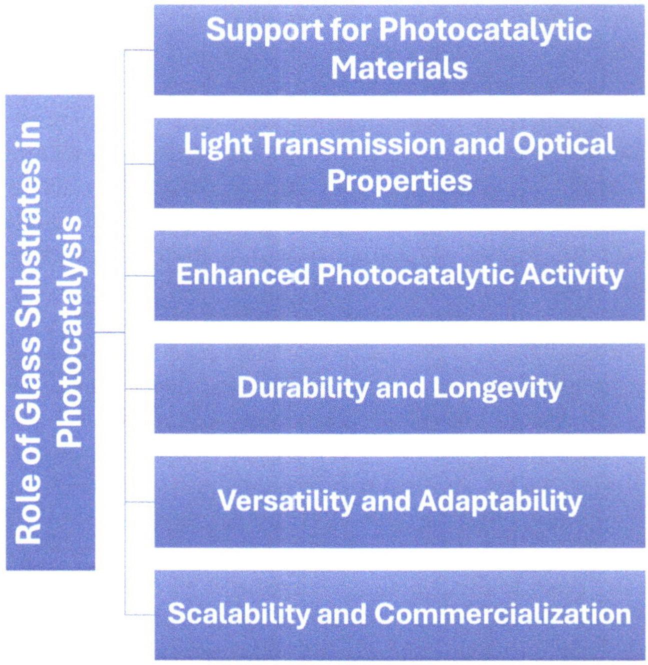

Figure 8.2. Role of glass substrates in photocatalysis.

their extreme adaptability. To accommodate various photocatalytic materials and reaction circumstances, they may be customized to take on certain forms, dimensions, and surface functions. Glass substrates can be utilised in a variety of configurations to meet the needs of distinct photocatalytic processes and reactor designs, such as flat sheets, fibres, membranes, and coatings. Because glass substrates are so versatile, tailored photocatalytic systems may be created for a variety of industrial, energy-producing, and environmental remediation applications [12]. Glass substrates make photocatalytic technologies more commercially viable and easier to use by providing scalability and compatibility with large-scale production processes. Glass-based photocatalytic devices may be made with commonly accessible materials and inexpensive production methods. Glass substrates' scalability makes it possible to produce photocatalytic coatings and devices in large quantities for use in air purification systems, industrial manufacturing facilities, and municipal water treatment plants. Glass substrates offer an economical and useful way to put photocatalysis into practice on a commercial level. The special qualities of glass substrates guarantee the long-term efficiency and dependability of photocatalytic systems, improve light transmission, and encourage uniform coating deposition [9]. Glass-based photocatalytic technologies have enormous promise for tackling global concerns linked to resource conservation, energy sustainability, and environmental pollution with further study and technical innovation.

8.3.3 Significant achievements in photocatalyst coating on glass substrates

Over the past few decades, the development of photocatalyst coatings on glass substrates has achieved major milestones, fueled by the rising need for sustainable solutions in industrial applications, energy production, and environmental remediation. These significant developments in surface engineering, materials science, and photocatalytic technology open up new avenues for the use of photocatalytic coatings on glass substrates in real-world applications. The photocatalytic characteristics of titanium dioxide (TiO_2), a semiconductor substance that may catalyze chemical processes when exposed to light, were discovered by researchers in the late 20th century. The development of photocatalytic coatings for environmental cleanup and energy generation applications was made possible by this groundbreaking discovery. The creation of self-cleaning glass surfaces was one of the first uses of photocatalyst coatings on glass substrates [12–16]. Researchers were the first to apply TiO_2 coatings to building facades and glass windows in the 1990s in order to produce self-cleaning surfaces that could degrade organic contaminants and stop dirt buildup. This invention provided a long-term solution for keeping surfaces sanitary and clean, revolutionizing the automotive and architectural sectors. Significant progress has been achieved in photocatalyst deposition methods for coating glass substrates throughout the years. To create consistent and sticky photocatalytic coatings on glass surfaces, methods including electrodeposition, sol–gel deposition, CVD, and spray pyrolysis have been developed. By precisely controlling the coating's thickness, shape, and composition, these deposition processes improve the coated glass substrates' photocatalytic performance and durability. The production of nanostructured photocatalysts with improved characteristics and performance has been made possible by the development of nanotechnology, which has completely changed the area of photocatalysis.

High surface area-to-volume ratios and distinctive optical and electrical characteristics make nanomaterials such as nanoparticles, nanotubes, and nanowires extremely useful for photocatalytic applications. To increase photocatalytic activity and stability, researchers have looked at incorporating nanostructured photocatalysts into coatings for glass substrates. By adjusting the characteristics of photocatalytic materials and coating deposition conditions, researchers have concentrated on improving the photocatalytic performance of coatings on glass substrates [16–18]. To improve light absorption, charge separation, and reaction kinetics in photocatalytic coatings, techniques including heterostructure creation, surface modification, and metal ion doping have been used. The creation of extremely effective and selective photocatalytic coatings for a range of applications is the result of these optimization efforts. Photocatalyst coatings on glass substrates have been shown to have a wide range of practical uses, such as energy production, air filtration, water purification, environmental remediation, and self-cleaning surfaces. Photocatalytic coatings have been used in consumer goods, industrial production facilities, and municipal water treatment plants on a commercial and pilot scale, demonstrating the viability and scalability of this technology. Multifunctional coatings, which combine photocatalytic activity with other characteristics including antibacterial qualities, anti-fogging capabilities, and UV protection, are the result of recent advances in materials science and surface engineering. The range and influence of photocatalyst coatings on glass substrates are increased by these multifunctional coatings, which provide improved performance and adaptability for a variety of applications.

8.4 Methods of applying photocatalysts onto glass substrates

Techniques for coating glass substrates with photocatalysts are essential for a number of applications, such as solar panels, air purification systems, and self-cleaning glass. Figure 8.3 provides list of methods used for making photocatalysts on glass substrates for multipurpose applications [13–19]. A number of methods, each with pros and cons, have been devised to deposit photocatalytic compounds onto glass surfaces in an efficient and consistent manner.

8.4.1 Sol–gel process

In order to create a stable network, the sol–gel process first synthesises a sol, or a colloidal suspension of nanoparticles in a liquid precursor, which is then gelated. Glass substrates are commonly coated or dipped in a sol containing photocatalytic nanoparticles, including zinc oxide (ZnO) or titanium dioxide (TiO_2), when it comes to photocatalytic coatings. A thick and adherent photocatalytic layer forms on the glass surface more easily with further heat treatment, which is depicted in figure 8.3(a). The sol–gel technique is appropriate for large-scale manufacturing

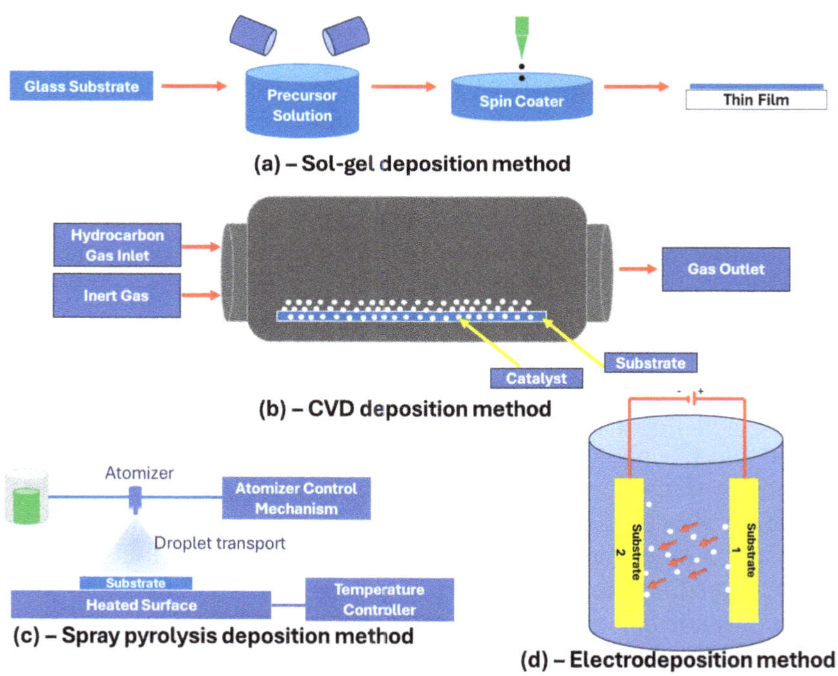

Figure 8.3. Deposition methods for making photocatalysts on glass substrates.

and a variety of applications because it provides exceptional control over film thickness, content, and shape. The basic mechanism of the sol–gel process is the regulated hydrolysis and condensation processes that convert a precursor solution (sol) into a three-dimensional network (gel) [10–13]. The aforementioned procedure may be customized to deposit precise thin films onto substrates, rendering it perfect for the creation of photocatalytic coatings on glass. The capacity of the sol–gel process to manufacture coatings with exact control over thickness, content, and structure is one of its main benefits. This degree of control is essential for maximizing photocatalyst performance. Researchers can modify precursor concentration, pH, and drying conditions to customize the properties of the final coatings to fit particular application needs. For example, altering the concentration of the precursor can have an impact on the coating's thickness, which in turn has an impact on the coated glass substrate's durability and photocatalytic activity. In addition, the sol–gel method provides superior adhesive qualities, guaranteeing that the photocatalytic coatings stick to the glass substrate with strength. Even in the face of extreme climatic conditions, the coated surfaces' lifetime and durability are increased by this tight link. Hence, in comparison to traditional coating methods, glass substrates coated with the sol–gel approach show improved resistance to deterioration, weathering, and abrasion [16]. The sol–gel method's compatibility with a variety of photocatalytic materials and doping agents is another noteworthy benefit. Due to its exceptional stability and photocatalytic activity, titanium dioxide (TiO_2) is still one of the most widely used photocatalysts. However, researchers may alter its optical and electrical characteristics by adding dopants, such as metals or non-metals. For instance, doping TiO_2 with nitrogen can increase its photocatalytic effectiveness when exposed to solar radiation by expanding its light absorption range into the visible spectrum. The consistent characteristics of the coated glass substrate are ensured by the homogenous dispersion of dopants inside the coating matrix, made possible by the sol–gel technique. Moreover, the sol–gel method is economical and scalable, which makes it appropriate for producing photocatalytic glass coatings on a wide scale. The sol–gel approach may be used with very basic equipment and normal laboratory facilities, unlike vapor-phase deposition procedures, which call for specialized equipment and regulated settings. Because of its accessibility, it is a desirable choice for industrial applications where scalability and cost-effectiveness are critical. The sol–gel method of applying photocatalytic coatings onto glass substrates has a wide range of applications in many industries and sectors. Through the process of photocatalytic oxidation, photocatalytic glass coatings have the potential to break down organic pollutants, eradicate hazardous bacteria, and cleanse air and water in the context of environmental remediation. The improvement of indoor air quality and the prevention of the transmission of infectious illnesses in public areas are two main advantages of these coatings [17]. Additionally, photocatalytic glass coatings aid in the production of surfaces that are self-cleaning and resist dust, dirt, and organic residues in architectural design and construction. Building facades, windows, and solar panels may all retain their visual appeal and performance with little upkeep by utilizing the photocatalytic qualities of these coatings.

8.4.2 Chemical vapor deposition (CVD)

One of the most popular methods for depositing photocatalysts onto glass substrates is the CVD process, which has several advantages in a range of applications such as energy generation, environmental remediation, and advanced materials research. Utilizing gas phase chemical processes, CVD provides accurate control over the composition, thickness, and homogeneity of the coating. This makes it a perfect option for depositing thin films of photocatalytic materials like titanium dioxide (TiO_2) onto glass surfaces. Precursor gases are essentially introduced into a reaction chamber during the CVD process, where they undergo chemical reactions and break down to generate a solid coating on the substrate surface which is depicted in figure 8.3(b). High-quality photocatalytic coatings with specific features may be produced more easily by fine-tuning the deposition conditions through the selection of precursor gases and process parameters [17–19]. The capability of the CVD process to produce conformal and homogeneous coatings on intricately geometrized glass surfaces is one of its main benefits. This capacity extends to complicated and three-dimensional substrates. Contrary to other deposition methods that could have trouble with uneven coating coverage or imperfections, CVD makes sure that a uniform layer of photocatalytic material is applied to the whole surface of the glass substrate. To maximize the photocatalytic efficacy and dependability of the coated surfaces, this homogeneity is necessary. Moreover, the CVD method provides superior control over the deposited films' microstructure and crystallinity, producing coatings with ideal photocatalytic qualities. Researchers may control the growth kinetics and crystallographic orientation of the deposited material, improving its catalytic activity and stability across a range of working circumstances, by varying factors including temperature, pressure, and precursor flow rates. The adaptability of the CVD process in depositing a broad range of photocatalytic materials and compositions onto glass substrates is another noteworthy benefit. Because of its stability and effectiveness, titanium dioxide is still one of the most researched photocatalysts. However, CVD enables the deposition of other materials, including zinc oxide (ZnO), tungsten oxide (WO_3), and several metal oxides and nitrides [20]. Because of its adaptability, scientists may investigate new photocatalytic systems with features specifically designed for certain uses, such as pollutant degradation and visible-light photocatalysis.

Additionally, the CVD technique allows for fine control over the doping of foreign elements into photocatalytic materials, expanding their application and improving their photocatalytic efficacy. The optical, electrical, and surface characteristics of the coated films may be changed by researchers by adding dopants, such as metal ions, nitrogen, or sulfur, during the deposition process. This improves light absorption, charge separation, and reaction kinetics. Apart from its technical benefits, the CVD technique provides repeatability and scalability, which makes it appropriate for photocatalytic glass coatings manufacturing on an industrial scale [19–21]. Large-scale production operations' throughput and quality needs can be satisfied by customizing CVD systems through appropriate equipment design and process parameter optimization. Because of its scalability, CVD is a good option for incorporating photocatalytic coatings into a wide range of goods and uses, such as environmental sensors, photovoltaic devices, and architectural glass and building materials.

8.4.3 Spray pyrolysis

Spray pyrolysis is a useful approach that offers numerous benefits for putting photocatalysts on glass substrates. These applications include environmental cleanup, renewable energy, and functional coatings. Spray pyrolysis is a regulated chemical method that enables the deposition of thin coatings of photocatalytic compounds, including titanium dioxide (TiO_2), onto glass surfaces by breaking down precursor solutions. The simplicity and scalability of the spray pyrolysis process are two of its main features. Spray pyrolysis, in contrast to more intricate deposition methods, entails atomizing precursor solutions into tiny droplets that are subsequently sprayed onto the substrate surface using a spray nozzle [20, 21]. The Spray pyrolysis process's adaptability and simplicity allow photocatalytic coatings to be quickly and affordably deposited over huge regions of glass substrates, which is depicted in figure 8.3(c). Because of its scalability, spray pyrolysis is especially well-suited for industrial-scale manufacturing, where cost-effectiveness and high through-put are crucial factors. Moreover, the spray pyrolysis technique provides superior control over the deposited films' composition, shape, and thickness. Researchers can modify the precursor concentration, spray flow rate, substrate temperature, and deposition duration to satisfy certain performance needs while also customizing the coatings' characteristics. The density and crystallinity of the photocatalytic nano-particles, for example, may be influenced by varying the precursor concentration and deposition temperature. This can have an impact on the stability and photo-catalytic activity of the coated glass substrate. The ability of the spray pyrolysis process to produce homogeneous and conformal coatings over intricate and irregularly shaped substrates is another important benefit. Precursor droplets are uniformly dispersed throughout the substrate surface thanks to the atomization and spray deposition processes, producing coatings with constant thickness and cover-age. For the coated glass substrates to function reliably and maximize photocatalytic performance, this homogeneity is essential, especially in situations where exact control over the coating qualities is critical. Moreover, a variety of photocatalytic materials and compositions may be deposited onto glass substrates with adaptability using the spray pyrolysis approach. Researchers can investigate other metal oxides and sulfides, zinc oxide (ZnO), tungsten oxide (WO_3), and titanium dioxide as alternative photocatalysts. Because of its adaptability, customized photocatalytic coatings with improved performance qualities such as visible-light absorption, charge carrier mobility, and catalytic reactivity can be created [22]. In summary, the spray pyrolysis technique offers simplicity, scalability, homogeneity, and diversity, making it a viable and adaptable method for putting photocatalysts onto glass substrates. Researchers and industry may investigate new avenues for using photocatalytic coatings in a variety of applications, from solar energy harvesting and functional material creation to environmental remediation and air purification, by capitalizing on the special benefits of this technology.

8.4.4 Electrodeposition

Applying an electric field to a solution containing metal ions or precursor molecules allows for the electrodeposition of thin coatings onto conductive surfaces by electrochemical deposition. When it comes to photocatalytic coatings on glass substrates, electrochemical processes enable the deposition of a thin layer of photocatalytic material onto the glass surface when conductive coatings or substrates are submerged in a solution containing photocatalytic precursors. It is possible to produce coatings on surfaces with complicated shapes using electrochemical deposition, which also allows for perfect control over layer composition and thickness. Its application is limited to certain glass substrates and combinations, though, as it needs conductive substrates or coatings. In order to create a solid layer on the substrate surface, electrochemical deposition fundamentally uses an electric field to accelerate the reduction or oxidation of precursor ions found in an electrolyte solution [17, 19]. This method can be used for surface functionalization, energy harvesting, environmental remediation, and other applications where homogenous and conformal coatings have to be deposited across vast regions of glass substrates. The capacity to precisely regulate coating thickness, content, and morphology is one of the main benefits of the electrochemical deposition technique. Researchers may modify the electrolyte content, deposition duration, and deposition potential to satisfy particular performance requirements while also changing the properties of the deposited films, which is depicted in figure 8.3(d). By optimizing photocatalytic activity, light absorption, and charge transfer kinetics, this degree of control raises the coated glass substrates' total efficiency. Moreover, a variety of photocatalytic materials and compositions may be deposited onto glass substrates via electrochemical deposition, thanks to its adaptability. Researchers can investigate other metal oxides and sulphides, ZnO, WO_3, and titanium dioxide as alternative photocatalysts. Because of their adaptability, customized photocatalytic coatings with improved performance qualities such as visible-light absorption, photoresponse, and catalytic reactivity can be created. The electrochemical deposition method's ability to work with intricate and three-dimensional substrate geometries is another noteworthy benefit. Electrochemical deposition is able to cover glass substrates with complex forms and features more successfully than other deposition processes that could have trouble with coating imperfections or non-planar surfaces [5, 13]. The potential uses of photocatalytic coatings in microfluidic devices, sensors, and optoelectronic devices are increased by this capacity. Moreover, the photocatalytic coatings will cling securely to the glass substrate surface because of electrochemical deposition's superior adhesion capabilities. Even in the face of extreme climatic conditions, the coated surfaces' lifetime and durability are increased by this tight link. Because of this improved resistance to deterioration, corrosion, and abrasion, glass substrates coated via electrochemical deposition are appropriate for demanding applications and prolonged outdoor exposure. The electrochemical deposition process has many technological benefits, but it also has financial and environmental benefits. It uses few consumables and generates little waste. A wide range of researchers and companies can use electrochemical deposition since it can be accomplished using very simple and affordable setups, unlike certain physical vapor deposition processes that may need vacuum systems and specialized equipment [18]. Researchers and industry may investigate new avenues for using photocatalytic coatings in many applications, such as energy harvesting, advanced materials science, surface engineering, and environmental remediation, by leveraging the distinct benefits of this approach.

8.5 Applications of photocatalyst-glass composites

A family of materials known as photocatalyst-glass composites incorporates photocatalytic nanoparticles such as ZnO or TiO_2 into glass matrices to provide new functions and characteristics. These composites, which make use of the special combination of photocatalytic activity and transparency that glass substrates provide, find a wide range of applications in different sectors.

8.5.1 Self-cleaning glass

The photocatalytic activity of the coating material, which permits the breakdown of dirt and grime and the destruction of organic pollutants when exposed to ultraviolet (UV) or visible light, is the fundamental idea underlying photocatalyst-glass composites. Reactive oxygen species (ROS), including hydroxyl radicals, are produced by the photocatalyst during irradiation. These radicals have potent oxidative qualities and can degrade organic molecules that have been adsorbed on the glass surface [19–21]. The ability of photocatalyst-glass composites to provide self-cleaning qualities without the need for harsh chemicals or manual cleaning procedures is one of its main advantages. All it takes to start a continuous cleaning cycle and remove organic pollutants from the coated glass surface is exposure to natural sunshine or artificial light sources. This process keeps the surface clean throughout time. In addition, photocatalyst-glass composites are more resilient and long-lasting than traditional surface treatments or coatings. On the glass substrate, the photocatalytic coating creates a strong, chemically stable layer that offers long-term defence against environmental elements including moisture, UV rays, and temperature changes. This guarantees that even under difficult external circumstances, the self-cleaning capability will continue to work. The potential of photocatalyst-glass composites to enhance the visual appeal of glass surfaces by reducing the accumulation of dirt, water spots, and streaks is another noteworthy advantage [22]. For architectural applications where keeping a neat and appealing look is crucial, such as building facades, windows, and skylights, this makes them especially well-suited. Furthermore, by lowering the need for chemical and water cleansers, which can be harmful to ecosystems and public health, photocatalyst-glass composites promote environmental sustainability. Through the use of solar energy to initiate the self-cleaning process, these composite materials provide an environmentally responsible way to keep surfaces clean and sanitary while reducing their negative effects on the environment [23–26]. To sum up, photocatalyst-glass composites are a potential new development in self-cleaning glass applications; they provide effective, long-lasting, and eco-friendly solutions for a variety of sectors

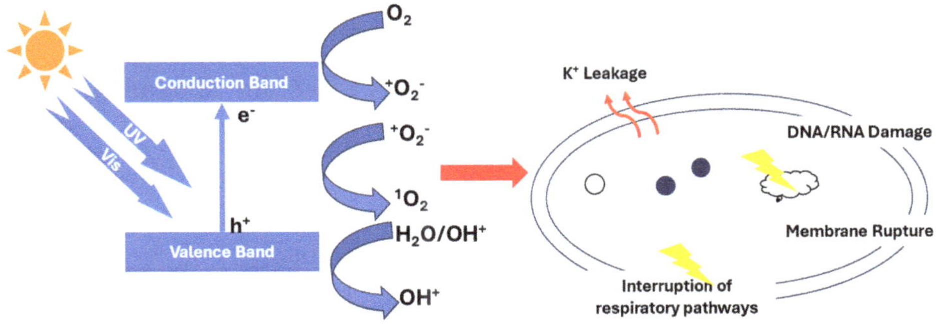

Figure 8.4. The photocatalytic self-cleaning antibacterial action's proposed mechanism.

and uses. These composites allow for the construction of self-cleaning surfaces that enhance hygiene and cleanliness while also supporting sustainability and energy efficiency in the built environment by incorporating photocatalytic coatings onto glass substrates.

Numerous theories have been put out to explain why TiO_2 photocatalyst has antimicrobial properties. When TiO_2 is exposed to appropriate light, a number of ROS are produced, including superoxide, hydrogen peroxide, and hydroxyl radical. These ROS have the ability to be lethal to the microorganisms, as shown in figure 8.4. Bacterial cells can have their cell walls and membranes destroyed by TiO_2 radiation [24]. According to certain studies, photogenerated holes created by exposure to visible light do not have the necessary reduction capability to make radicals .OH radical as a result of H_2O oxidation. Less oxidative O_2 radicals, such as $^1O_2$ are assumed to be in charge of the photocatalytic breakdown of bacterial cells when exposed to visible light [25–27]. But stimulation by UV radiation can also result in the production of highly oxidizing radicals .OH radicals, which have the ability to inactivate microorganisms by photocatalysis.

8.5.2 Air purification and systems

In air purification systems, photocatalyst-glass composites are a cutting-edge technology that provide a practical and long-lasting means of eliminating dangerous contaminants and enhancing indoor air quality. These composites are made up of glass substrates covered in a thin coating of a photocatalytic substance, usually TiO_2, which uses light to catalyze surface chemical processes. When exposed to light, the coating material's photocatalytic activity facilitates the oxidation and breakdown of airborne contaminants, including microorganisms, nitrogen oxides (NO_x), and VOCs [17, 22]. When exposed to UV or visible light, the photocatalyst produces ROS, including hydroxyl radicals. These radicals have potent oxidative qualities that enable them to degrade organic and inorganic pollutants into innocuous byproducts like CO_2 and water. The capacity of photocatalyst-glass composites to remove pollutants constantly and efficiently from interior spaces without the need for extra filtering or chemical treatments is one of their main benefits in air purification systems.

Figure 8.5 depicts in daylight, the three primary uses of TiO_2-modified building materials. Similar functionalities might be shown for interior air applications, with the exception of the need for more lighting in a space. Contaminants are in touch with the photocatalytic coating on the glass surface as air flows through the environment, where they degrade and change into innocuous chemicals. By lowering the number of contaminants in the air, this procedure serves to make interior environments healthier and more breathable for residents. In addition, photocatalyst-glass composites provide a number of advantages over conventional air purification techniques, such as increased durability, less maintenance needs, and environmental friendliness [13–18]. On the glass substrate, the photocatalytic coating creates a strong, chemically stable layer that can endure extended exposure to light and air. In air purification applications, this guarantees long-term performance and dependability, minimizing the need for regular replacement or service. Furthermore, photocatalyst-glass composites support environmental sustainability by using artificial or natural light sources to power air purification. Photocatalyst-glass composites use renewable energy sources and do not create waste or hazardous byproducts, in contrast to typical air purifiers that depend on power or replaceable filters. They are therefore a sustainable and environmentally beneficial way to enhance indoor air quality while lowering negative effects on the environment [22]. These composites facilitate the creation of sustainable and affordable air purification

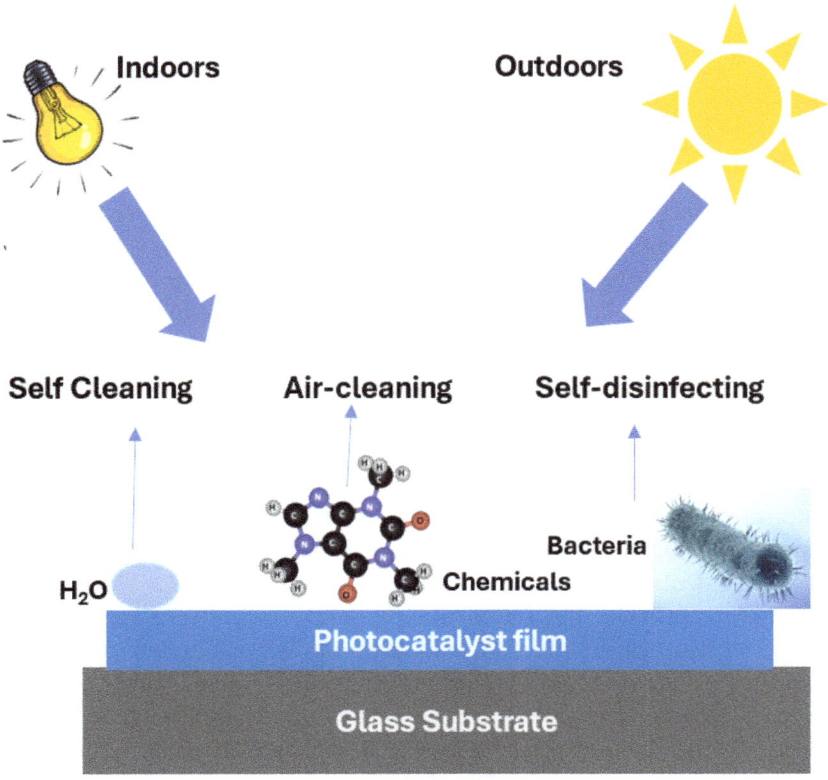

Figure 8.5. Graphic depiction of a multipurpose photocatalytic construction material.

solutions for industrial, commercial, and residential settings by incorporating photocatalytic coatings onto glass substrates.

8.5.3 Anti-microbial surfaces

A novel approach to creating antimicrobial surfaces, photocatalyst-glass composites provide durable and efficient defence against a variety of bacteria, viruses, and other diseases. These composites are made of glass substrates covered in a thin coating of photocatalytic material, usually TiO_2, which, when triggered by light, has potent antibacterial characteristics. When exposed to light, the coating material's photocatalytic activity produces ROS, which have potent oxidative qualities that can harm and even kill microorganisms. Examples of ROS include hydroxyl radicals and superoxide ions. Microorganisms are inactivated and eventually eliminated when they come into touch with the photocatalyst-glass surface because of the antimicrobial properties of ROS [23]. The capacity of photocatalyst-glass composites to offer constant, passive defence against microbial contamination is one of their main benefits in antimicrobial surface applications. Photocatalyst-glass surfaces, in contrast to conventional disinfection techniques, which depend on chemical agents or labour-intensive cleaning techniques, constantly function to prevent the development, and spread of germs and viruses without requiring human intervention [24–26]. This makes them especially ideal for high-touch surfaces in public places, medical institutions, and regions where hygienic conditions are crucial. Additionally, compared to traditional antimicrobial coatings, photocatalyst-glass composites have a number of benefits, such as increased longevity, minimal maintenance needs, and broad-spectrum action against a range of pathogens.

Under low light conditions, titanium dioxide (TiO_2) nanocomposite antimicrobial coatings are one of the best ways to rid common surfaces of pathogens (bacteria, fungus, and viruses), which is depicted in figure 8.6. It shows the photocatalytic disinfection effectiveness of newer TiO_2 nanocomposite antimicrobial coatings for surfaces, dental, and orthopedic implants. TiO_2 is primarily combined with inorganic metals (such as copper (Cu), silver (Ag), manganese (Mn), etc), non-metals (such as fluorine (F), calcium (Ca), and phosphorus (P)), and two-dimensional materials (such as MXenes, MOF, and graphdiyne) to control the charge transfer mechanism, surface porosity, crystallinity, and the effectiveness of microbial disinfection [25, 27]. The titanium (Ti) metal implants are coated with silane functionalizing chemicals and polymers to create superhydrophobic properties that prevent bacteria adherence. TiO_2 nanocomposite coatings have been shown to exhibit remarkable biocorrosion resistance, endurance, biocompatibility, bone-formation potential, and long-term antibacterial efficacy in dental and orthopedic metal implants.

On the glass substrate, the photocatalytic coating creates a strong, chemically stable layer that can endure multiple washing and disinfection cycles without losing its antimicrobial effectiveness. This lowers the chance of microbial contamination and transmission by ensuring long-term efficacy and dependability in antimicrobial treatments. Additionally, by using artificial or natural light sources to drive antimicrobial action, photocatalyst-glass composites support environmental

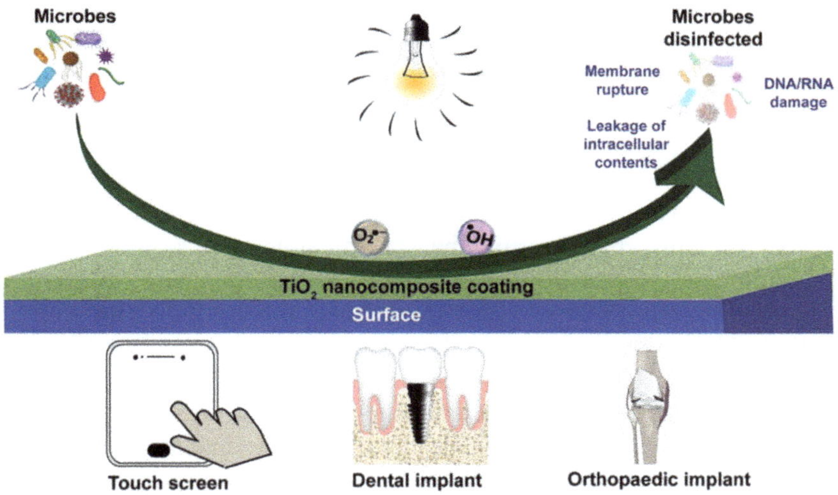

Figure 8.6. Antimicrobial TiO_2 nanocomposite coating for multipurpose applications. Reprinted from [27] with permission from Elsevier, CC BY 4.0.

sustainability [28]. In contrast to chemical disinfectants, which could be hazardous to both the environment and human health, photocatalytic coatings work using renewable energy sources and leave no residues or dangerous byproducts behind. As a result, they provide an environmentally safe and low-impact way to create antimicrobial surfaces. These composites allow the creation of sustainable and affordable antimicrobial solutions for a variety of applications, such as the food processing, hotel, and healthcare sectors, by incorporating photocatalytic coatings onto glass substrates.

8.5.4 UV protection applications

Innovative solutions for UV protection applications are provided by photocatalyst-glass composites, which shield against damaging ultraviolet (UV) radiation effectively while preserving transparency and clarity. These composites are made of glass substrates covered in a thin coating of a substance that is photocatalytic such as TiO_2 that has the ability to both absorb and deflect ultraviolet light. When exposed to UV light, the coating material's photocatalytic activity allows for the production of ROS, including hydroxyl radicals. By absorbing and diffusing the energy of damaging UV photons, these ROS molecules efficiently neutralise them and reduce the amount of UV radiation that penetrates the glass substrate [22–26]. Consequently, photocatalyst-glass composites aid in shielding both indoor and outdoor spaces from the harmful effects of ultraviolet light, such as sunburn, early ageing, and skin cancer. The ability of photocatalyst-glass composites to offer broad-spectrum UV filtering without sacrificing optical clarity or transparency is one of their main benefits in UV protection applications. In contrast to conventional UV-blocking films or coatings that could change the color or look of glass surfaces, photocatalyst-glass composites provide excellent UV protection without sacrificing the aesthetic appeal of glass. This makes them perfect for applications requiring clarity and visibility, including as car windows, architectural glazing, and eyeglasses. In addition, photocatalyst-glass composites are more resilient and long-lasting than traditional UV protection techniques. On the glass substrate, the photocatalytic coating creates a strong, chemically stable layer that is resistant to deterioration and loss of functionality even after extended exposure to UV light and other environmental variables [27]. In UV protection applications, this guarantees long-term efficacy and dependability, minimizing the need for regular replacement or maintenance. Additionally, by lowering the transmission of UV light through glass surfaces, photocatalyst-glass composites assist to minimize energy consumption by regulating interior temperatures and minimizing the need for air conditioning systems [28]. These composites limit heat gain and enhance thermal comfort in interior areas by blocking UV rays, which may result in energy savings and a decrease in carbon emissions. These composites enable the creation of UV-protected environments in a variety of industries and sectors, including architecture, automotive, and eyewear, encouraging sustainability, comfort, and health. This is achieved by embedding photocatalytic coatings onto glass substrates.

8.5.5 Water treatment and purification

In water treatment and purification applications, photocatalyst-glass composites are a revolutionary breakthrough that provide effective and long-lasting solutions for eliminating impurities and enhancing water quality. When exposed to UV or visible light, the coating material's photocatalytic activity allows for the formation of ROS, including hydroxyl radicals and superoxide ions. These ROS molecules have potent oxidative qualities that enable them to break down inorganic impurities found in water as well as organic pollutants and microbiological infections [29]. The ability of photocatalyst-glass composites to provide continuous and efficient purification without the need for additional chemicals or filtering systems is one of their main advantages in water treatment applications. In figure 8.7(a), a standard slurry reactor system comprises a reactor that holds suspended TiO_2 photocatalysts, an LP Hg UV light source enclosed in quartz, and membrane filtration for catalyst recovery. Figures 8.7(b)–(d) represent more contemporary reactor designs that do not require membrane separation after treatment by using immobilized catalyst particles. Contaminants are exposed to the oxidative and antibacterial properties of ROS when water passes over the photocatalytic coating on the glass surface, which causes their breakdown and elimination from the water matrix. Cleaner and safer water is produced for a variety of uses by this procedure, which also helps to get rid of organic compounds, heavy metals, pesticides, viruses, bacteria, and other dangerous microorganisms [27–29]. In addition, photocatalyst-glass composites provide a number of advantages over conventional water treatment techniques, such as increased durability, less maintenance needs, and environmental friendliness. On the glass substrate, the photocatalytic coating creates a strong, chemically stable layer that is resistant to deterioration and loss of functionality even after extended exposure to water and UV light. In water purification applications, this guarantees long-term performance and dependability, minimizing the need for

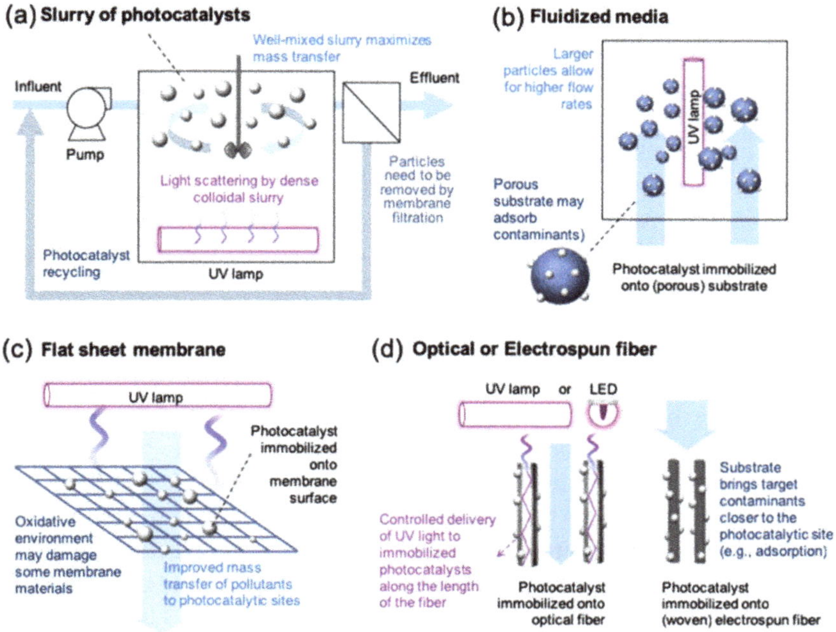

Figure 8.7. Reactor designs for water treatment using semiconductor photocatalysis. Reprinted with permission from [29], copyright {2018} American Chemical Society.

regular replacement or service. Additionally, by using artificial or natural light sources to power water purification, photocatalyst-glass composites support environmental sustainability. Photocatalytic coatings work with renewable energy sources and do not produce secondary pollutants, in contrast to chemical disinfectants or filtering systems that may. They are therefore a green and environmentally beneficial way to cleanse and purify water with the least possible negative effects on the environment. Photocatalyst-glass composites have the potential to be used for wastewater treatment and polluted water body rehabilitation in addition to their purifying properties [29]. Photocatalytic coatings can be used to decrease pollution, lower microbiological contamination, and enhance the general safety and health of water resources by being applied to water treatment infrastructure, such as filter units, pipelines, and reservoirs. These composites allow for the integration of photocatalytic coatings onto glass substrates, thereby facilitating the development of environmentally friendly and economically viable water purification systems for a range of industries and sectors, such as point-of-use water filtration, industrial wastewater management, and municipal water treatment.

8.6 Challenges and future directions

8.6.1 Durability and longevity of photocatalyst coatings

Glass substrates coated with photocatalysts provide a viable option for a range of applications, including UV protection, air purification, and self-cleaning surfaces.

Evaluating these coatings' endurance and durability is essential to determining their efficacy and suitability for various real-world applications. First off, a number of variables, including the calibre of the coating procedure, the selection of the photocatalytic material, and the surrounding environment, affect how long photocatalyst coatings on glass substrates last [19–22]. Superior deposition techniques including electrochemical deposition, spray pyrolysis, and CVD provide homogeneous and adherent coatings, which increases their endurance. To stop delamination or deterioration over time, the coating's adherence to the glass substrate is crucial. Furthermore, the choice of photocatalytic substance has a big impact on how long the coatings last. Materials with high stability and resistance to deterioration in severe environments and under UV radiation include TiO_2. But new developments in material science have made photocatalysts like ZnO and WO_3 that are more durable and have higher photocatalytic activity possible. The lifespan of photocatalyst coatings can be impacted by environmental conditions such as exposure to sunshine, moisture, temperature fluctuations, and chemical contaminants. The photocatalytic process may be triggered by UV light, which can result in the breakdown of organic contaminants and self-cleaning properties. Long-term UV exposure, however, may potentially result in surface fouling or photocatalyst deterioration, which would shorten the coating's lifespan. The development of mineral deposits or water spots on the glass surface can be attributed to moisture and humidity levels, which may eventually lessen the coating's efficacy. The lifespan and robustness of photocatalyst coatings on glass substrates can be increased by regular cleaning and appropriate maintenance. Frequent cleaning promotes ideal light exposure and photocatalytic activity by removing accumulated dust, grime, and organic residues [22–27]. Applying surface treatments or protective coatings can also increase the photocatalyst coatings' resilience to environmental stresses and lengthen their lifespan. Researchers and industry can create durable photocatalyst coatings for a variety of applications and help create cleaner, safer, and more sustainable surroundings by addressing these aspects and using advancements in material science and coating technology.

8.6.2 Scale-up and commercialization challenges

Photocatalyst coatings on glass substrates have promising uses in a variety of sectors but scaling them up and commercializing them presents considerable obstacles. A number of significant obstacles must be overcome before moving on to large-scale manufacture and commercial deployment, even though laboratory-scale production may show viability and effectiveness. A significant obstacle in the process of scaling up is attaining a uniform and consistent coating deposition over sizable glass surfaces. Many deposition processes used in laboratories, such CVD, and sol–gel procedures, could not be easily scaled up to satisfy the production requirements of commercial applications. It takes process parameter, equipment design, and deposition technique optimization to scale up these operations without sacrificing coating quality and uniformity [28]. Moreover, scale-up and commercialization initiatives heavily depend on the selection of photocatalytic material and coating formulation. Because of its stability and photocatalytic activity, titanium dioxide (TiO_2) is still the most often used photocatalyst; nevertheless, new materials with better performance attributes are always being developed. The identification of scalable and economically viable production processes for these materials is important for their commercial viability.

Making sure photocatalyst coatings on glass substrates are compatible with current production techniques and materials is another difficulty in commercializing them. The integration of architectural fabrication techniques or established glass production lines necessitates a meticulous evaluation of substrate compatibility, adhesive qualities, and coating endurance. For broad use, it is imperative to develop coating formulations that cling well to different glass compositions and surface treatments while retaining optical clarity and transparency. Furthermore, meeting safety and regulatory criteria presents serious difficulties for commercialization [19–22]. The use of chemical additives or nanoparticles in photocatalyst coatings raises questions about their potential effects on the environment, human health, and workplace safety. It is imperative to adhere to regulatory norms and requirements for product safety, emissions, and disposal in order to secure market approval and maintain customer trust. Furthermore, two important variables affecting the economic feasibility of photocatalyst coatings are cost-effectiveness and market competitiveness. In commercialization efforts, striking a balance between performance, durability, and production costs to fulfil end-users' pricing expectations while offering greater functionality is crucial. Researchers, manufacturers, regulators, and end users must work together to create scalable manufacturing methods, optimize coating formulas, guarantee product safety, and boost market competitiveness in order to overcome these obstacles. Notwithstanding the challenges, photocatalyst coatings have enormous potential to improve environmental quality, promote sustainability, and improve human health and well-being.

8.6.3 Integration with smart glass technologies

The combination of smart glass technologies with photocatalyst coatings on glass substrates is a major development in the field of environmentally sustainable building materials. Smart glass, often referred to as dynamic glass or switchable glass, is made using technologies that enable it to alter its characteristics in reaction to outside stimuli like heat, light, or electricity. Smart glass can provide improved functionality and performance for a range of applications when paired with photocatalyst coatings [28]. A significant benefit of combining photocatalyst coatings with smart glass technologies is the creation of surfaces that are capable of actively eliminating organic pollutants, dirt, and grime. Photocatalytic materials, including TiO_2 utilize light's catalytic properties to facilitate chemical processes that break down organic molecules and eliminate microorganisms. Photocatalyst coatings, when used on smart glass, allow the surface to self-clean in reaction to natural or artificial light, minimizing the need for routine maintenance and cleaning [30]. Additionally, photocatalyst coatings combined with smart glass technologies can improve building comfort and energy efficiency. As the environment changes, smart glass may dynamically change its transparency or reflectivity, which can assist control of inside temperatures, lessen glare, and cut down on the need for artificial lighting and HVAC systems. Smart glass can also help with air purification and odour reduction by combining photocatalyst coatings, resulting in healthier and cosier interior spaces. In addition, the integration of photocatalyst coatings with smart glass technologies has prospects for enhanced customization and functionality [30]. For instance, photocatalyst coatings may be engineered to precisely regulate the optical characteristics and performance of glass by selectively absorbing or reflecting particular light wavelengths [15–19]. Smart glass may adjust in real time to user preferences, ambient factors, and energy needs through integration with automation systems and sensor technologies, improving user comfort and productivity. Smart glass with photocatalyst coatings is a cutting-edge approach to sustainable development, green construction, and modern architecture because it combines energy-efficient features, air-purifying capabilities, and self-cleaning qualities. The combination of photocatalyst coatings with smart glass technologies is set to completely change how we interact with and perceive glass surfaces in the built environment as long as research and innovation in this sector continue to progress.

8.7 Conclusion

To sum everything up, investigating the technical developments of photocatalysts on glass substrates opens up a world of opportunities and breakthroughs that have significant ramifications for a variety of sectors and applications. Gaining an understanding of photocatalysis's fundamentals reveals how important it is for solving environmental problems and developing sustainable practices. From architecture to healthcare, photocatalysis is important in many industries because it provides solutions for environmental remediation, antimicrobial surfaces, and air and water purification. Glass substrate characteristics are closely linked to photocatalysis, highlighting their critical function as the basis for photocatalyst coatings. Glass substrates improve the efficacy and efficiency of photocatalytic processes in addition to offering stability and longevity. In material science and engineering, photocatalyst coating on glass substrates has reached important milestones. These groundbreaking discoveries, which ranged from innovative research to large-scale production, set the stage for revolutionary applications in environmental remediation, air and water purification, and self-cleaning surfaces. Photocatalysts may be applied on glass substrates in a variety of ways that provide flexibility and accuracy, allowing customized solutions for a range of applications. These techniques, which range from CVD to electrochemical deposition, enable companies and researchers to maximize the performance and attributes of coatings. Photocatalyst-glass composites have a broad range of applications, including improving indoor air quality, reducing water pollution, and inhibiting the development of microorganisms. These composites show how sustainability and technology may work together to create cleaner, healthier surroundings. Challenges like endurance and durability, scaling up, and commercialization highlight the necessity of ongoing research and innovation. To fully realize the promise of photocatalyst coatings on glass substrates, multidisciplinary collaboration and strategic collaborations are necessary to overcome these obstacles. An exciting new age of intelligent surfaces with adaptive features and improved performance is heralded by integration with smart glass technologies. Through the integration of dynamic glass qualities with the advantages of photocatalysis, creative solutions for sustainable development, improved user experience, and energy-efficient structures are generated. Essentially, the development of photocatalyst coatings on glass substrates represents a paradigm change towards more sustainable, healthier, and cleaner living conditions. The path towards a more promising and environmentally conscious future is still being forged as we successfully traverse the obstacles and seize the possibilities that lie ahead.

Acknowledgments

GSDB is grateful to the Management of Chettinad College of Engineering and Technology, Karur for constant support and encouragement. Swetha Madamala acknowledges the Management of MVJ College of Engineering for their constant support.

References

[1] Yang X and Wang D 2022 Photocatalysis: from fundamental principles to materials and applications *ACS Appl. Energy Mater.* **1** 6657–93

[2] Byrne C, Subramanian G and Pillai S C 2018 Recent advances in photocatalysis for environmental applications *J. Environ. Chem. Eng.* **6** 3531–55

[3] Kubacka A, Fernandez-Garcia M and Colon G 2012 Advanced nanoarchitectures for solar photocatalytic applications *Chem. Rev.* **112** 1555–614

[4] Ibhadon A O and Fitzpatrick P 1998 Heterogeneous photocatalysis: recent advances and applications *Catalysts* **3** 189–218

[5] Wenderich K and Mul G 2016 Methods, mechanism, and applications of photodeposition in photocatalysis: a review *Chem. Rev.* **116** 14587–619

[6] Ahmed S N and Haider W 2018 Heterogeneous photocatalysis and its potential applications in water and wastewater treatment: a review *Nanotechnology* **29** 342001

[7] Marcelino R B P and Amorim C C 2019 Towards visible-light photocatalysis for environmental applications: band-gap engineering versus photons absorption—a review *Environ. Sci. Pollut. Res.* **26** 4155–70

[8] Nair V, Muñoz-Batista M J, Fernández-García M, Luque R and Colmenares J C 2019 Thermo-photocatalysis: environmental and energy applications *ChemSusChem.* **12** 2098–116

[9] Lu M 2013 *Photocatalysis and Water Purification: from Fundamentals to Recent Applications* (New York: Wiley)

[10] Humayun M, Raziq F, Khan A and Luo W 2018 Modification strategies of TiO_2 for potential applications in photocatalysis: a critical review *Green Chem. Lett. Rev.* **11** 86–102

[11] Dunnill C W and Parkin I P 2011 Nitrogen-doped TiO_2 thin films: photocatalytic applications for healthcare environments *Dalton Trans.* **40** 1635–40

[12] Rtimi S, Dionysiou D D, Pillai S C and Kiwi J 2019 Advances in catalytic/photocatalytic bacterial inactivation by nano Ag and Cu coated surfaces and medical devices *Appl. Catal.* B **240** 291–318

[13] Shwetharani R, Chandan H R, Sakar M, Balakrishna G R, Reddy K R and Raghu A V 2020 Photocatalytic semiconductor thin films for hydrogen production and environmental applications *Int. J. Hydrogen Energy* **45** 18289–308

[14] Blizak D, Remli S, Blizak S, Bouchenak O and Yahiaoui K 2021 NiO thin films for environmental photocatalytic applications: a review *Alger. J. Environ. Sci. Technol.* **7** 1950–7

[15] Abdel-Galil A, Hussien M S A and Yahia I S 2021 Synthesis and optical analysis of nanostructured F-doped ZnO thin films by spray pyrolysis: transparent electrode for photocatalytic applications *Opt. Mater.* **114** 110894

[16] Sosnin I M, Vlassov S and Dorogin L M 2021 Application of polydimethylsiloxane in photocatalyst composite materials: a review *React. Funct. Polym.* **158** 104781

[17] Loyola Poul Raj I, Jegatha Christy A, David Prabu R, Chidhambaram N, Shkir M, AlFaify S and Khan. A 2020 Significance of Ni doping on structure-morphology-photoluminescence, optical and photocatalytic activity of CBD grown ZnO nanowires for opto-photocatalyst applications *Inorg. Chem. Commun.* **119** 108082

[18] Tismanar I, Obreja A C, Buiu O and Duta A 2021 VIS-active TiO_2–graphene oxide composite thin films for photocatalytic applications *Appl. Surf. Sci.* **538** 147833

[19] Najafidoust A, Allahyari S, Rahemi N and Tasbihi M 2022 Uniform coating of TiO_2 nanoparticles using biotemplates for photocatalytic wastewater treatment *Ceram. Int.* **46** 4707–19

[20] Liu J, Ye L, Sun Y, Hu M, Chen F, Wegner S, Mailänder V, Steffen W, Kappl M and Butt H-J 2020 Elastic superhydrophobic and photocatalytic active films used as blood repellent dressing *Adv. Mater.* **32** 1908008

[21] Jellal I, Nouneh K, Toura H, Boutamart M, Briche S, Naja J, Soucase B M and Touhami M E 2021 Enhanced photocatalytic activity of supported Cu-doped ZnO nanostructures prepared by SILAR method *Opt. Mater.* **111** 110669

[22] Lee Y, Fujimoto T, Yamanaka S and Kuga Y 2021 Evaluation of photocatalysis of Au supported ZnO prepared by the spray pyrolysis method *Adv. Powder Technol.* **32** 1619–26

[23] Dalawai S P, Aly Saad Aly M, Latthe S S, Xing R, Sutar R S, Nagappan S, Ha C-S, Kumar Sadasivuni K and Liu S 2020 Recent advances in durability of superhydrophobic self-cleaning technology: a critical review *Prog. Org. Coat.* **138** 105381

[24] Padmanabhan N T and John H 2020 Titanium dioxide based self-cleaning smart surfaces: a short review *J. Environ. Chem. Eng.* **8** 104211

[25] Pakdel E, Wang J, Kashi S, Sun L and Wang X 2020 Advances in photocatalytic self-cleaning, superhydrophobic and electromagnetic interference shielding textile treatments *Adv. Colloid Interface Sci.* **277** 102116

[26] Geyer F, D'Acunzi M, Sharifi-Aghili A, Saal A, Gao N, Kaltbeitzel A, Sloot T-F, Berger R, Butt H-J and Vollmer D 2020 When and how self-cleaning of superhydrophobic surfaces works *Sci. Adv.* **6** eaaw9727

[27] Kumaravel V, Nair K M, Mathew S, Bartlett J, Kennedy J E, Manning H G, Whelan B J, Leyland N S and Pillai S C 2021 Antimicrobial TiO_2 nanocomposite coatings for surfaces, dental and orthopaedic implants *Chem. Eng. J.* **416** 129071

[28] Gil F J, Padrós A, Manero J M, Aparicio C, Nilsson M and Planell J A 2002 Growth of bioactive surfaces on titanium and its alloys for orthopaedic and dental implants *Mater. Sci. Eng.* C **22** 53–60

[29] Loeb S K, Alvarez P J J, Brame J A, Cates E L, Choi W, Crittenden J, Dionysiou D D *et al* 2018 The technology horizon for photocatalytic water treatment: sunrise or sunset? *Environ. Sci. Technol.* **53** 2937–47
[30] Li H, Tu W, Zhou Y and Zou Z 2016 *Z*-Scheme photocatalytic systems for promoting photocatalytic performance: recent progress and future challenges *Adv. Sci.* **3** 1500389

Chapter 9

Graphene glass-ceramic-based materials for photocatalytic wastewater remediation

R V Tolentino-Hernandez, M S Ovando-Rocha and F Caballero-Briones

Glass-ceramics materials are commonly used in construction, kitchenware, electronics, medical, energy, optical applications, among others, due to their versatile properties such as high transparency, good formability, and tunable chemical compositions. Glass-ceramics materials have polycrystalline aggregates within the vitreous matrix, which vary from 0.5 up to 95% depending on the process temperature and the composition of the crystalline phases. The wide range of amorphous glass and crystalline phases for these materials lead unique tunable properties for new emerging applications. Recently, glass-ceramics have shown promising results in wastewater remediation through the addition of photocatalytic materials such as TiO_2, ZnO, $LiNbO_3$, etc. Also, the incorporation of graphene oxide and reduced graphene oxide has attracted great attention for water remediation applications because they possess a very high specific surface area, and can be further decorated with ceramic, photocatalytic materials, reducing electron recombination. This chapter aims to review the recent work on glass-ceramic materials intended for water remediation. The chapter reviews the materials, new green synthesis methods and performance, focusing on the incorporation of graphene oxide and reduced graphene oxide for enhancing photocatalytic and absorption properties of glass-ceramic systems.

9.1 Introduction

9.1.1 Wastewater: origins and prediction

Water is a natural and vital resource for humans; the access to clean and safe water, sanitation and hygiene is the most basic human need according to the United Nations (UN), and it is included in the Sixth Goal 'Clean Water and Sanitation' of the 17 Sustainable Development Goals (SDG). Water is also important for

doi:10.1088/978-0-7503-5904-7ch9

agriculture, and the industrial sector, for manufacturing, cooling, cleaning, and other processes, which produce large amounts of wastewater.

The untreated or insufficiently treated effluents from industrial, agricultural, mining and domestic sources alter the composition of water bodies, adding a wide range of pollutants such as textile, paint, leather, or compounds related to agricultural, mining activities, sewage, pesticides or fertilizers, energy use, and so on [1]. Municipal wastewater from residential, commercial, institutional, and recreational facilities combined with industrial effluents and stormwater also contribute to water that requires treatment [2] which includes organic, inorganic, and biological contaminants [3].

As a result of this environmental problem, organizations and governments look to improve water quality by reducing pollution, eliminating dumping, and minimizing release of hazardous chemicals and materials, increasing the proportion of treated wastewater, and substantially increasing recycling and safe reuse, as well as expanding international cooperation and capacity building support to water management in developing countries.

Scientists have been also working on controlling water pollution, by investigating wastewater composition, health and ecosystem impacts, and the effects of environmental variables in pollutant degradation, interactions and so on; but also on developing novel materials, cleaner production processes and wastewater cleaning.

On the other hand, according to the UN, the world's population has increased more than three times since the 20th century and it is expected to increase to 9.7 billion in 2050, creating pressure on water resources, including freshwater availability as well as wastewater treatment (WWT). Figure 9.1 presents the increase in wastewater production in different world areas, stressing the need for increasing and better WWT strategies.

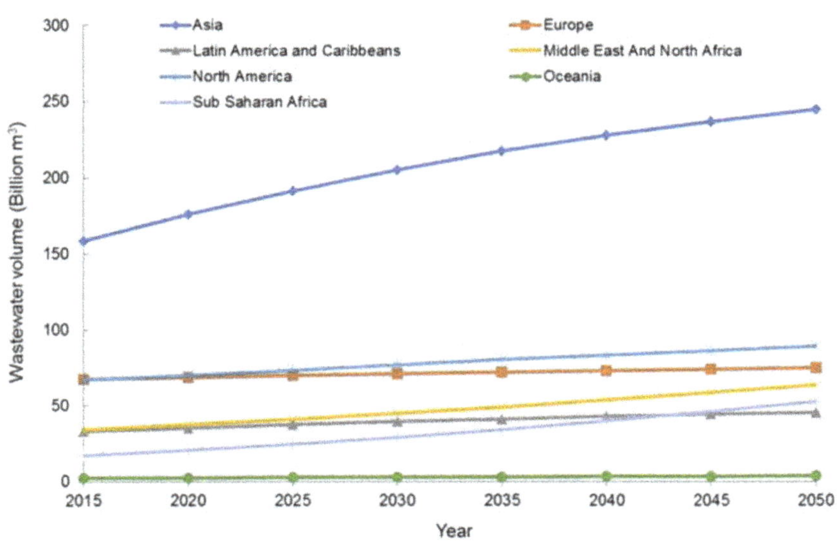

Figure 9.1. Wastewater production by regions in 2015 and predicted until 2050 [4] John Wiley & Sons. CC BY 4.0.

9.1.2 Wastewater treatments

Archeologists have found evidence of managed wastewater, among Nomadic tribes around 10 000 B.C.E, in Babylonia by 4000 B.C.E, in Indus valley circa 2500 B.C.E, and more, but it was not until the 6th century C.E. that the Romans developed and managed a modern wastewater system. Nowadays, with technology and scientific knowledge, WWTs has evolved [5].

Sewage, or WWT, is the action to remove pollutants present in water before it can be discharged to water bodies or reused in another process. WWT is a complex activity, which is often divided into five steps: (i) preliminary or pretreatment [6]; (ii) primary; (iii) secondary; (iv) tertiary; and (v) sludge treatment [7]. These steps may include mechanical (a series of tanks, pumps, screens, blowers or grinders [8]), physical (floating, flocculating, filtration, blending, screening and gas exchange), chemical (precipitation, sedimentation, coagulation, ion exchange, oxidation, neutralization and stabilization), biological (several kinds of separation aerobic and anaerobic treatment by microorganisms, aerated lagoons) [9], physicochemical (chemical flocculating- filtration, adsorptive bubbles separation techniques, ion flotation, solvent sublation-colloid flotation, photocatalysis and oxidation) [10] processes, as well as supervised dumping, recycling, or incineration of residues [11].

9.1.3 Emerging processes for wastewater treatment

Since the 1980s, the concerns on persistent and emerging pollutants such as pesticides, pharmaceuticals, plastics, and eutrophicating substances such as nitrogen and phosphorus species, to mention some [12], which are not adequately removed by traditional WWT [13], have led to novel approaches called advanced oxidation process (AOPs) [14, 15] or advanced oxidation technologies (AOTs) [16], which have been increasingly used in different academic and industrial media, as reflected in the numbers of publications in Scopus, as shown in figure 9.2.

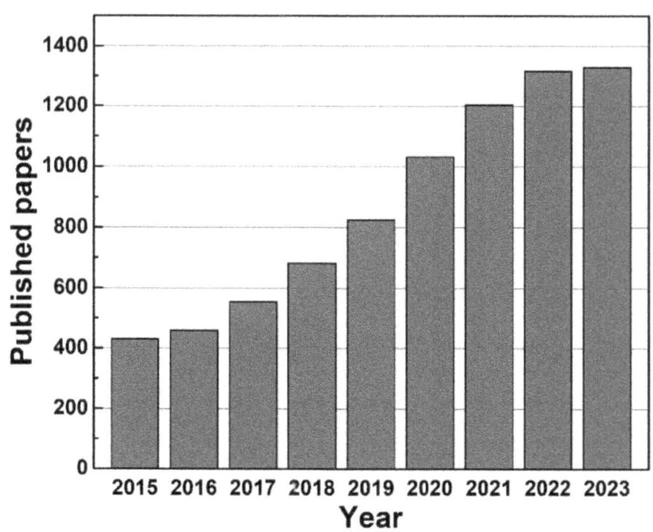

Figure 9.2. Research interest in the field of AOPs (retrieved from Scopus in October 2023).

9.2 Advanced oxidation processes

AOPs provide complete degradation of several organic or inorganic pollutants [17]. They are divided into homogeneous (HO-APSs) and heterogeneous (HE-AOPs), and subdivided into photochemical and chemical [18]. Figure 9.3 describes the classification of AOPs.

AOPs include photochemical degradation processes such UV/O_3, UV/H_2O_2, or photocatalysis, for example TiO_2/UV, photo-Fenton reactive [20], and chemical oxidation processes like O_3, O_3/H_2O_2, H_2O_2/Fe^{2+}, which refer to a set of chemical treatment procedures that degrade substances by oxidation through reaction with hydroxyl radicals (OH^-), which are very reactive, attacking most organic molecules, but they are not selective [21]. In these processes, the reaction rates are in the range of 10^8–10^{11} M^{-1} s^{-1}, the local concentration of OH^- radicals can be around 10^{10}–10^{12} M, and they have a high oxide potential ($E = 2.8$ V), giving these species a large effectivity to produce organic free radicals [22] (equation (9.1)).

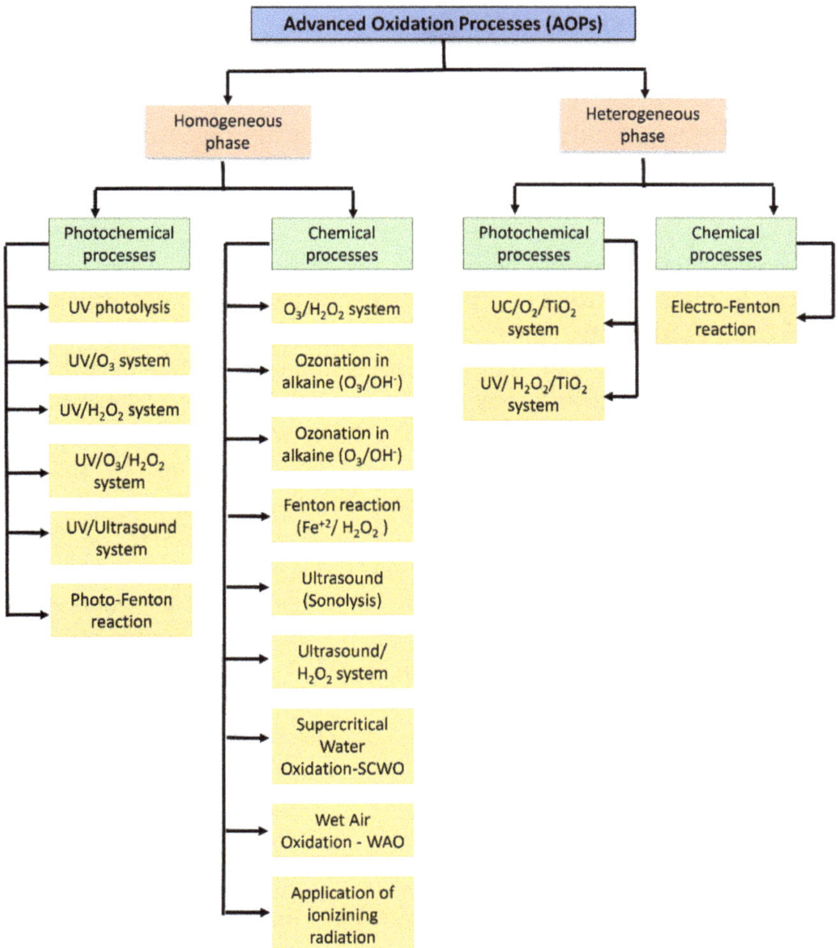

Figure 9.3. Types of AOPs employed by WWT. Adapted from [19] CC BY 4.0.

$$OH^0 + RH \rightarrow R^0 + H_2O \tag{9.1}$$

Some free organic radicals can react with molecular oxygen to obtain peroxiradicals and continue the reaction (equation (9.2)), which can start a series of oxidation reactions [23].

$$R^0 + O_2 \rightarrow RO_2^0 \rightarrow \text{products} + CO_2 \tag{9.2}$$

Some examples of hazardous and persistent contaminants are dinitrotoluene (DNT), trinitrotoluene (TNT), dimethyl sulfoxide (DMSO) or carbofuran (CBF) which have been degraded using AOPs, specifically with ozone and ultraviolet radiation through ozone photolysis; the general reactions are depicted below [18].

$$H_2O + O_3 \xrightarrow{h\nu} 2OH + O_2 \tag{9.3}$$

$$2OH \rightarrow H_2O_2 \tag{9.4}$$

AOPs decrease the chemical oxygen demand (COD) to ca. 70%–80% in contrast to 30%–45% of the traditional WWT [9].

9.2.1 Photocatalysis

Photocatalysis has been used to degrade a wide range of organic compounds such as dyes, harmful chemicals, and hydrocarbons from oil refinery wastewater, as it operates under mild conditions of temperature and sunlight or ultraviolet (UV) radiation [24].

The photocatalytic mechanism begins with the absorption of an incident photon in the photocatalyst; the electrons (e^-) in the valence band (VB) are promoted to the conduction band (CB) creating an electron–hole (e^-–h^+) pair also called exciton. The charge separation of the e–h pair into water, leads to the production of hydroxyl ($HO^\bullet$) or oxygen ($O_2^\bullet$) radicals, known as reactive oxygen species (ROS), which can oxidize the organic pollutants present in wastewater. These ROS are created on the surface of the photocatalyst by the reduction of the O_2 and H_2O molecules by the generated photoelectrons; in an O_2 excess, other ROS can be created such as hydrogen peroxide (H_2O_2), hydroperoxyl radical ($HO_2^\bullet$) and hydroperoxyl anion (HO_2^-). The presence and interaction of ROS, free electrons and holes in polluted water allows the degradation of diverse kind of contaminants like organic and inorganic matter, but also bacteria and viruses [25]. The scheme of the photocatalytic mechanism is shown in figure 9.4(a) and the ROS formation in figure 9.4(b).

The rate of the photocatalytic reaction depends on the intensity of the incident light, photocatalyst bandgap and thickness, as well as the rate of e^-–h^+ recombination [26, 27].

9.2.2 Semiconductor catalysts

The first investigations of photocatalysis were the photodecomposition of cyanide (CN^-) in water on TiO_2 surface, later of polychlorinated biphenyls ($C_{12}H_{10-x}Cl_x$) by TiO_2 under ultraviolet light irradiation. Subsequently, studies of halogenated

(a)

(b)

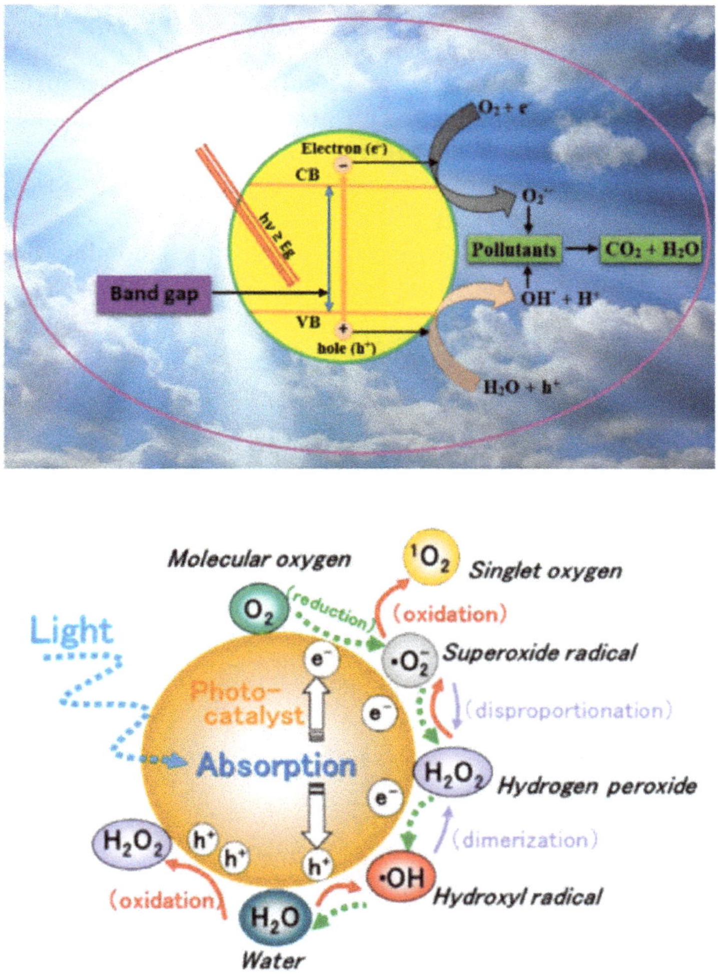

Figure 9.4. (a) Scheme of the photocatalytic mechanism, and (b) ROS formation mechanism. Reproduced from [25], Copyright (2023), with permission from Elsevier.

organic compounds formally proposed the oxidative decomposition behavior of semiconducting photocatalytic materials for organic pollutants.

Some semiconductors as metal transition oxides (TiO_2, ZnO, ZrO_2, Fe_2O_3, SnO_2, CeO_2, V_2O_5) or sulfides as CdS and ZnS are among the most popular heterogenous photocatalysts due to their high efficiency in the decomposition of organic matter, resulting in the formation of more biodegradable compounds or the total mineralization into carbon dioxide and water [28].

9.2.2.1 Transition metal oxides (TMOs)

Some examples of photocatalytic treatment using transition metal oxides are those from Dilaeyenna *et al* [29], using ZnO with polyvinylpyrrolidone (PVP), achieving a reduction of 90.61% of industrial dye wastewater from the newspaper printing industry; or that of Ganapathy *et al* [30] who worked with nano-sized ZnO biosynthesis from *Plectranthus amboinicus*, which led to a removal efficiency of ca. 93.75% [31]. Apart from TiO_2 and ZnO, other transition metal oxides used in photocatalytic treatment of polluted wastewater are CeO_2, ZrO_2, Fe_2O_3, SnO_2, NiO, Sb_2O_3, CuO or V_2O_5, as shown in figure 9.5.

Adding metal dopants to photocatalysts improve both light adsorption coefficient and photocatalytic degradation efficiencies. Increasing the charge separation efficiency, acting as a trapping site and delaying of the recombination of the excited electron are shown in figure 9.6 [32].

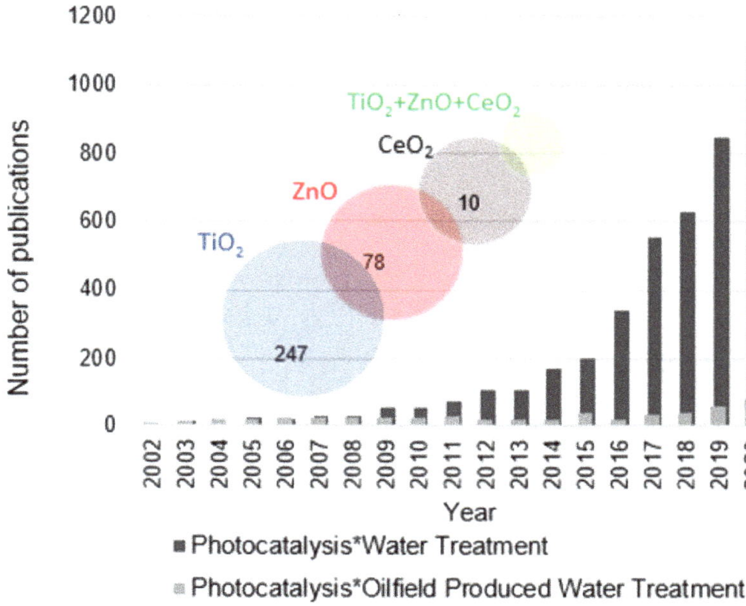

Figure 9.5. Publications on photocatalytic water treatment using metal oxides. Reproduced from [24] CC BY 4.0.

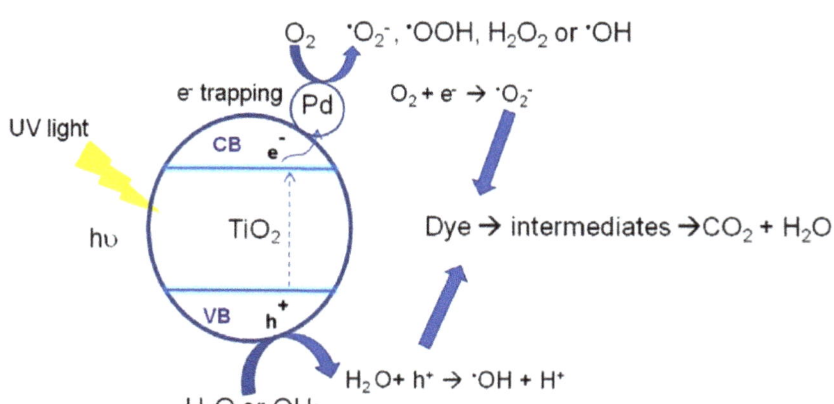

Figure 9.6. Example of doping as an electron trapping site of photocatalyst. Reproduced from [32] CC BY 4.0.

9.2.2.2 Transition metal chalcogenides (TMCs)

The chalcogenide group includes oxygen, sulfur, selenium, and tellurium. Transition metal chalcogenides (TMCs) have attracted interest because of their flexible elemental composition, tunable bandgap, Earth abundance, good optoelectronic properties, visible-light activity, and catalytic stability. TMOs and TMCs have been prepared through various methods to improve their crystallinity, and control the optical bad gap [33]. Some TMCs used for the photocatalytic oxidation of organic matter during WWT are CdS and ZnS [34].

9.3 New photocatalytic glass-ceramics materials for wastewater treatment

Glass exhibits optical transparency, hardness, impermeability to gases and liquids, good chemical resistance, and formability. On the other hand, ceramics are polycrystalline materials with several useful properties such as high-temperature resistance, high chemical inertness, good oxidation stability, lower density than metals, transparency, malleability, and high mechanical strength and hardness, among others [35]. Many ceramics have photocatalytic activity, as discussed above [36].

Thus, based on the combination of their respective properties, novel hybrid 'photocatalytic glass-ceramics' materials (PGCs) [37], have recently emerged for environmental applications such as air and water cleaning, self-cleaning and bacterial disinfection, offering the possibility of developing special microstructures with tuned properties [38], as shown in figure 9.7.

Photocatalytic glass-ceramics are classified in two main categories due to the involved crystallization mechanism: (i) volume (bulk), and (ii) surface crystallization, as shown in figure 9.8. When the dominant crystallization mechanism is volumetric, the value of the reduced glass transition temperature ($T_{gr} = T_g/T_m$) is below 0.58, this is because the temperature of crystal nucleation is lower than the

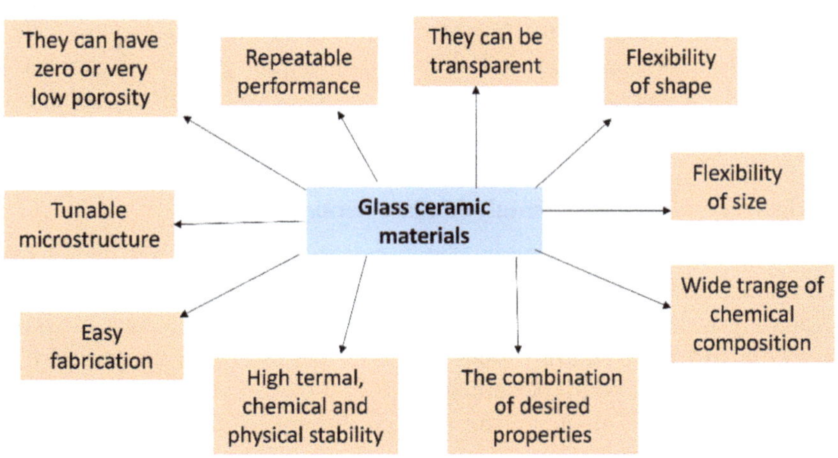

Figure 9.7. Properties of glass-ceramic materials.

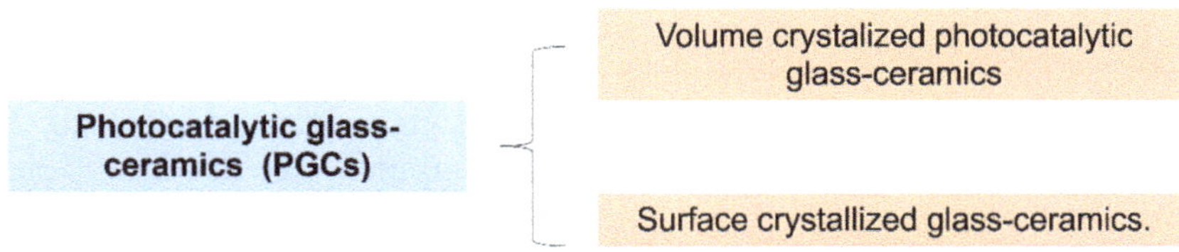

Figure 9.8. Photocatalytic glass-ceramics classification.

glass transition temperature (T_g) which if favorable for homogeneous nucleation; if $T_{rg} > 0.60$ the crystal nucleation becomes heterogeneous since the temperature nucleation is near to the T_g and surface crystallization mechanism dominates.

Another factor that may influence the type of crystallization mechanism is the density between the crystalline and glassy phases; if the difference is larger than 10%, the predominant mechanism is the surface one due the high elastic deformation energy; if it is less than 10%, the volume crystallization dominates the process. Even if some mechanism is dominant (volume or surface), the other one still happens in a minor proportion. The proper combination of both crystallization mechanisms may enhance the photocatalytic efficiency of the resultant material, coupled to the possibility of synthetizing photocatalysts with diverse shapes like rods, flowers, fibers, fuzzy grass, sheets, among others [39, 40].

9.3.1 Volume crystalized photocatalytic glass-ceramics

The volume crystallization involves the photocatalyst crystal nuclei growth in a random and homogeneous distribution through the entire volume of the glassy phase, hence, the microstructure of these PGCs is tightly interlinked with crystals from 50 to 100 nm up to 1 μm [38]. At the beginning, the nucleating agents form seeds which start the nuclei, then a spontaneous heterogeneous crystal growth happens, starting at the surface. Then, the parameters mentioned above rule the subsequent volume or surface crystallization process. After nucleation, the parent glass is processed at high temperatures to induce the crystal growth [37]. Other methodologie,s such as milling and shaping the parent glass and ceramic photocatalysts crystals before starting the sintering process, are commonly used, as shown in figure 9.9.

9.3.2 Surface crystalized photocatalytic glass-ceramics

Most glass-ceramics take advantage of volumetric nucleation, but there are glasses in which a controlled crystallization allows surface nucleation, which is best suited for photocatalysis, a phenomenon related to the surface; thus, surface crystallization is more significant for obtaining photocatalytic glass-ceramic materials with high photocatalytic activity and higher efficiency [41]. For surface crystallization growth, additional to the T_{rg} control, or the density difference between the ceramic and vitreous phase, some physical or chemical methods can be implemented to enhance the surface growth of the photocatalytic ceramic, since the crystal growth can occur

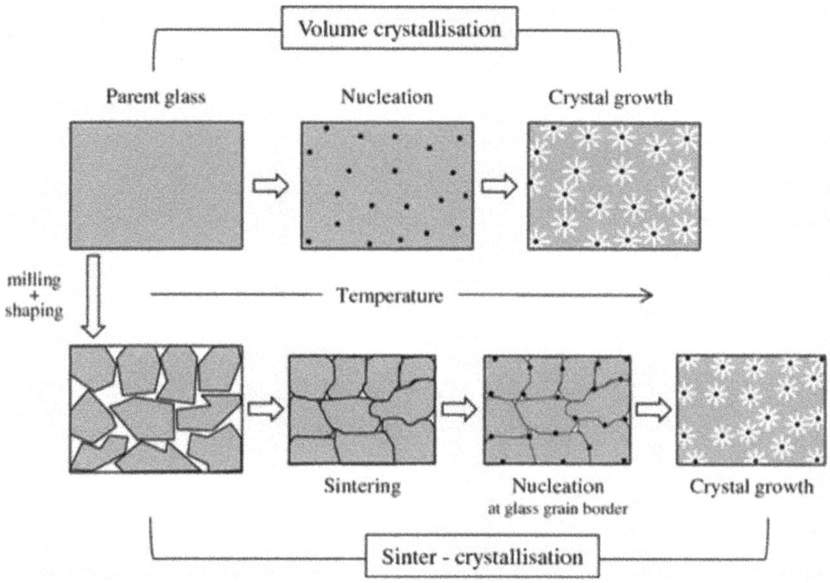

Figure 9.9. Volume crystallization of glass-ceramics. Reprinted with permission from Springer Nature [38].

within glass to the borders, although the preferential growth is towards inside [37, 39]. Figure 9.10 shows the surface crystallization process, where a $CaO–Al_2O_3–SiO_2$ glass was granulated and sintered at 850 °C without the addition of nucleation agent, which causes the crystal nucleation at the interface of the granules; after a heating post-treatment at 1100 °C, crystals of B-wollastonite grow perpendicular to the interface with a needle shape over the surface [37].

Yazawa *et al* [42], managed to synthesize rutile-type TiO_2 glass-ceramic in a $SiO_2–Al_2O_3–B_2O_3–CaO–TiO_2$ glass system, in a temperature range of 973 K to 1173 K; to improve the photocatalyst activity, an acid leaching by HNO_3 was performed after the crystallization process: this allows obtaining a porous PGC which improves the photocatalytic efficiency four times for methylene blue dye degradation, compared with a rutile-like TiO_2-coated photocatalyst obtained by sol–gel method, due possibly to the increase of the surface area by the leaching.

By applying heat treatment, transparent nanocrystalline glass-ceramic materials with photocatalytic activity can be obtained, which makes them promising materials for WWT, as proven by Margha *et al* [43], who reported the synthesis of photocatalytic glass-ceramic materials based on ZnO and $TiO_2–ZnO$ in a glassy matrix of $SiO_2–B_2O_3–Na_2O–K_2O–P_2O_5–Li_2O–BaO$ by conventional melting at 1200 °C by 2–3 h. After that, a thermal treatment at 450 °C during 10 h was applied for the photocatalyst conversion at the glass-ceramic surface and the surface modification into a porous one. The photocatalytic degradation efficiency of the prepared materials was tested in real surface water and wastewater polluted with humic acid (HA) and dyes. $TiO_2–ZnO$ glass-ceramic photocatalyst showed the best photocatalytic performance with a 53% degradation of HA in the surface water

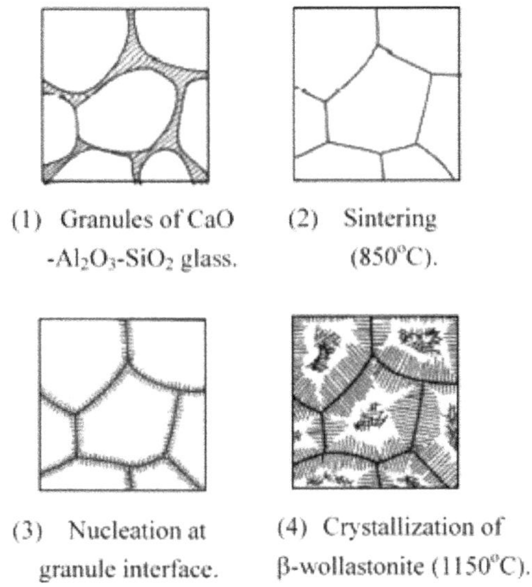

Figure 9.10. Sintering crystallization for the production of surface crystallization glass-ceramic [37]. John Wiley & Sons. Copyright (2010).

sample; in the wastewater sample, a 70% of decolorization and 60% remotion of chemical oxygen demand was achieved. Authors pointed out that the glass-ceramics photocatalysts can be reused several times with a slight decrease in the removal of organic pollutants.

Another example is the work presented by Gaur *et al*, where a glass-ceramic with a composition of $20.1Na_2O–23.1BaO–23TiO_2–17.4SiO_2–7.6B_2O_3–5.8Fe_2O_3–3Al_2O_3$ was synthesized by melt quench process at 1500 °C; then heat treatments at 650 °C for 2 h (HT-2), 4 h (HT-4), or 8 h (HT-8) were done. However, although a more ordered phase of $BaTiO_3$ was found in samples HT-8 and HT-4, sample HT-2 presented higher more methylene blue degradation activity. Sample HT-2 has more surface area and therefore more active sites were available to degrade methylene blue dye. Moreover, authors showed that this glass composition is favorable for piezocatalytic activity and can be used for water cleaning applications, in addition to showing moderate antibacterial properties in disinfecting *Escherichia coli* [44].

For the optimal operation of supported photocatalysts in wastewater purification some general requirements must be fulfilled, such as chemical stability, low toxicity, low water solubility, hydrophobicity, transparency, water percolation, high surface area, design shape, good thermal and mechanical properties, etc, as shown in figure 9.11.

Most of the requirements for supported photocatalytic materials (SPM) can be obtained in a new design of glass-ceramic composites, with the use of the very extensive list of photocatalysts, glass-ceramic matrixes and graphene-related materials for the tuning of the bandgap, percolation threshold, mechanical properties, chemical stability, and targeting the specific type of pollutants to be degraded.

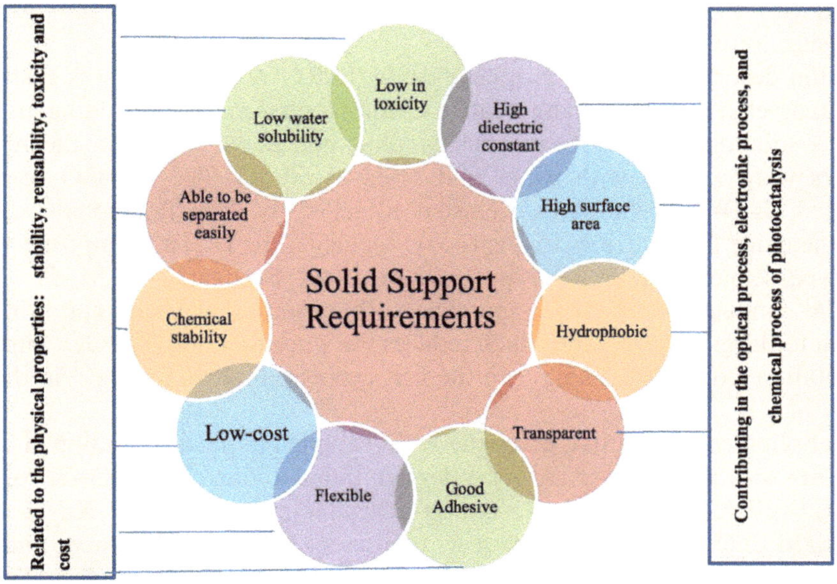

Figure 9.11. General requirements for a supported photocatalyst in wastewater remediation. Reproduced from [25], Copyright (2023), with permission from Elsevier.

9.3.3 New advanced photocatalyst/graphene glass-ceramics in wastewater treatment

The relatively low effectiveness of photocatalytic materials compared for example, with AOPs, has led to new proposals to increase the photocatalytic activity of semiconductor photocatalysts, in addition to doping or co-sorbents. One of these proposals is the incorporation to the photocatalytic materials different carbon allotropes such as carbon nanotubes [45], graphene and graphene oxide [46], and carbon nitride [47], to mention some. These compounds have improved charge separation and carrier capabilities, a wider light absorption range, and high dye adsorption capacity [48].

Some of these graphene-semiconductor photocatalysts have been widely used for pollutant degradation, or photocatalytic hydrogen synthesis [48]. For example, Gnanaseelan *et al* [49], explored the photocatalytic hydrogen generation in two ceramic-graphene composites, MoS_2–TiO_2–rGO (MTG) and CeO_2–$Ce_2Ti_3O_{8.7}$–TiO_2–rGO (CTG), synthetized by the hydrothermal method doing an alkaline pretreatment to commercial TiO_2 nanoparticles to produce a titanate interlayer and induce stoichiometry defects. Oxygen vacancies (O_v) and Ti^{3+} cation were observed in the $KTi_8O_{16.5}$, which created interband states below the conduction band which act as photoelectron traps, reducing charge recombination. The $Ce_2Ti_3O_{8.7}$ interlayer in CTG produce the same O_v and Ti^{3+}, but also mixed Ce^{4+}/Ce^{3+} cations which reduce the optical bandgap of the composite creating more interband states, enhancing optical absorption and improving the catalytic activity. Graphene drives the charge separation of the composites and acts as co-catalyst in the hydrogen photogeneration. The H_2 evolution of water/methanol with UV illumination of

254 nm was 363.83 μmol g^{-1} h^{-1} for MTG and 355.9 μmol g^{-1} h^{-1} for CTG composites, respectively.

Cesium-decorated reduced graphene oxide (CsRGO) composite was synthetized by Gnanaseelan *et al* [50], using a facile titration-based technique adding an CsOH aqueous solution into a GO suspension until pH 12.34, and then the resultant powders were washed with water and dried at 50 °C. The CsRGO composite presented Cs_2O-like nanoparticle clusters attached to some functional groups of graphene oxide like hydroxyl and carboxyl replacing the H. The composite showed an n-type semiconductor behavior, with a direct bandgap of 2.85 eV and an electrical conductivity of 10^{-4} S cm. The authors state that Cs doping brings one electron to the graphene plane which reduces the work function of RGO, improving H_2 evolution from 1701 μmolg^{-1} in the first cycle up to 2000 μmolg^{-1} in the third one.

Other photocatalysts using graphene or GO applied to the degradation of organic pollutants with a variety of reported degradation activities are ZnO-G, Ag/AgCl/GO, Ag/AgBr/GO, $ZnFe_2O_4$-G, among other [48]. For example, Xu *et al* [51], synthetized ZnO/graphene composite with different loadings of graphene and tested their photocatalytic activity in the photodegradation of MB under UV light, reaching a reaction rate constant k of 0.098 min^{-1} for the graphene load of 2 wt. % which is higher compared with that the pure ZnO ($k = 0.022$ min^{-1}). Zhu *et al* [52] synthetized Ag/AgCl/GO and Ag/AgBr/GO composites via a water/oil system at room temperature, the composites exhibit a visible-light-driven plasmonic photocatalytic activity, due to the hybridization of Ag/AgCl and Ag/AgBr with GO which enhance the electron charge transfer, suppressing the electron–hole pair recombination, improving the photocatalytic activity of the composites in the degradation test of methyl orange under visible-light irradiation. Fu *et al* [53], synthetized magnetic $ZnFe_2O_4$-graphene ($ZnFe_2O_4$-G) nanocomposite photocatalysts with different amounts of graphene by one-pot hydrothermal method. The photocatalytic efficiency of $ZnFe_2O_4$-G was higher than the pure $ZnFe_2O_4$; according to the authors, the GO used in the reaction was reduced during the hydrothermal process, then the sp^2 character of pristine graphene was restored enabling the charge carriers to behave as massless fermions due to the zero bandgap semiconductor properties of graphene, which improves the charge separation of the generated photoelectron in the $ZnFe_2O_4$ conduction band and transfer to the graphene sheet at the instant that they formed. The photocatalytic performance was proven in the photoelectrochemical degradation of MB to be almost 100% and for the photoelectrochemical decomposition by the generation of strong oxidant hydroxyl radicals, with the added benefit that the photocatalyst powders can be magnetically recovered. Another example is an experiment that analyzed the photocatalytic activity of graphene compounds, which reported that it is possible to degrade 100% of methylene blue dye using Ag/rGO; the authors suggested that this nanocatalyst does not present a risk in water treatment, being an excellent option for the elimination of organic contaminants from wastewater [54].

As mentioned above, the photocatalytic performance of different ceramic photocatalysts with graphene is enhanced by the increase in the photoelectron charge separation provided by the high electrical conductivity of graphene. That is why the incorporation of graphene/photocatalyst composite in a glassy matrix could be a way to create new SPMs, tailored to the needs for *in situ* water body treatments or in the creation of WWT plants. However, in the revised literature only few reports of glass-ceramic composites with the incorporation of graphene were found, where graphene is mostly used as a nanofiller to improve the glass-ceramic mechanical [55], tribological [56], electromagnetic interface shielding [57], or electrical properties [58].

As far as the completion of this chapter, there are no reports of the synthesis of photocatalytic glass-ceramic composites with the addition of graphene as nanofiller, either of graphene/photocatalysts embedded in a glassy matrix or of its use in tests of photocatalytic degradation of pollutants in water. This is why it could be interesting to investigate the photocatalytic behavior of embedded graphene/semiconductor in a glassy matrix for water remediation. The following section describes the methods used to produce graphene glass-ceramics which can be tuned to the obtention of supported photocatalytic graphene glass-ceramics with better photodegradation activity.

9.3.4 Synthesis of photocatalytic graphene glass-ceramic composites (P-GGCCs)

Photocatalysts consisting of transition metal oxide or chalcogenides decorating graphene oxide or reduced graphene oxide, dramatically expand the photocatalytic activity because of the better photocatalyst dispersion within the GO/rGO matrix, increased active area and improved charge separation, as discussed above. Therefore, it would be desirable for the preparation of graphene-based glass-ceramic photocatalysts to take advantage of the combined properties of glass and graphene-boosted photocatalytic ceramics.

However, a major challenge in preparing GGCCs is the development of synthesis routes to produce a good dispersion of graphene flakes into the glass-ceramic matrix, since the dispersion degree of graphene directly affects the final properties of the composite [59]. The common synthesis methods for graphene powders are two top-down methods: high energy milling (HEM) [60], and oxidation of graphite into graphene oxide by Hummer's method (HM) [61]. Although both methods produce graphene with structural defects, they have a high production yield compared for example, with chemical vapor deposition. However, HEM has a lack of quality control over the synthesis process and unexfoliated graphite, and non-homogeneous flakes and crystallite sizes are obtained in the powders. Correspondingly, in HM the graphene flakes are damaged by the strong oxidation process, introducing a lot of defects and decreasing the physical properties of the obtained graphene powders, which in turn can negatively affect the properties of the obtained GGCC [59]. Defect-free graphene can also be obtained by techniques such as liquid phase exfoliation (LPE), ultrasonication or electrochemical exfoliation [62, 63]. Nevertheless, graphene oxide is more suitable to the production of photocatalytic composites due to its functional groups that allow graphene to serve as a platform to anchor photocatalyst nanoparticles, in addition Hummer's method can be scaled at industrial level.

As mentioned above, the dispersion of graphene powders is the key point for the obtaining good quality and properties of GGCC. Typical powder dispersion methods are, sol–gel processing, powders dispersion by ultrasonication and HEM, colloidal processing, polymer-derived ceramics, etc. The classical dispersion by powder processing uses a combination of ultrasonic and HEM methods to obtain well-dispersed slurries of the GGCC, as shown in figure 9.12(a); this processing allows working with different glass-ceramic matrixes such as borosilicate glass, silica, silicon nitride, alumina, and zirconia [64].

Colloidal processing is another technique to produce GGCC in the basis of colloidal chemistry, where a mixture of colloidal suspensions of graphene and glass-ceramic powders use the same solvent to obtain an homogeneous dispersion for both materials, assisted by ultrasonication or simple magnetic stirring, although in some cases, a surface modification of the matrix or fillers is required, which can be done by the oxidation of graphene during HM, or by the use of surfactants for the generation of opposite charges between GC matrix and graphene [65], as schematized in figure 9.12(b).

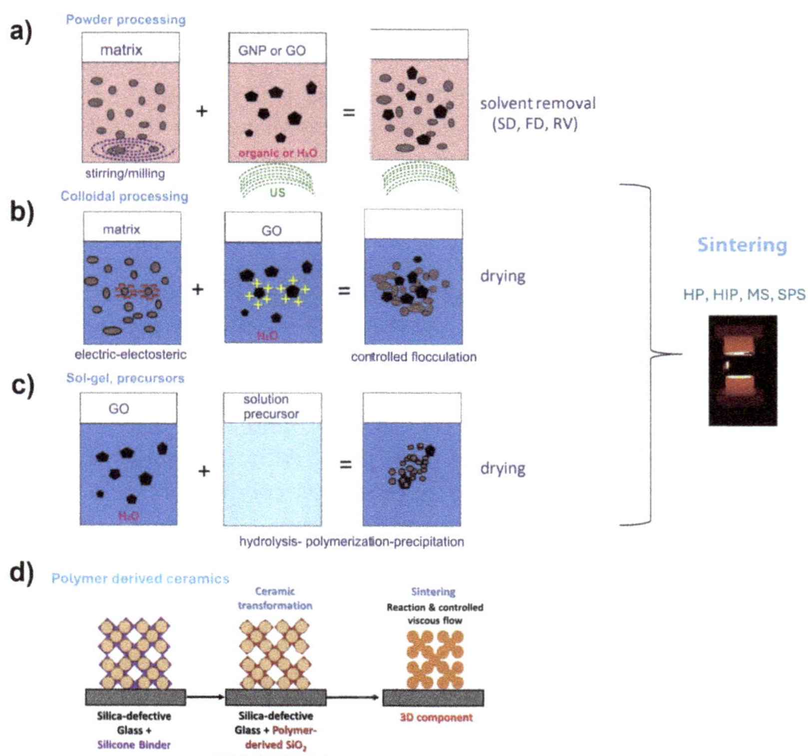

Figure 9.12. P-GGCC processing routes. (a) Powder processing, (b) colloidal processing, and (c) sol–gel processing. Reproduced from [65]. Copyright (2017), with permission from Elsevier and (d) polymer derived ceramics. Reproduced from [69]. Copyright (2020), with permission from Elsevier.

Sol–gel uses precursors which condensate to produce a solid phase with well-dispersed graphene flakes; for example, Ilyas *et al* [66], synthesized by a sol–gel route a graphene oxide-reinforced bioactive glass-ceramics composite, using $Ca(OH)_2$, TEOS and $(NH_4)_2HPO_4$ as source of glassy matrix components (CaO–SiO_2–P_2O_5) and graphene, taking advantage of the Ca^{2+} from glassy phase and COO^- groups of GO to interact and anchor to the graphene sheets, as shown in figure 9.12(c). Another advantage of the sol–gel process [67] is the potential to obtain transparent thin films and bulk materials with stable mechanical properties, as well as the dispersion in the liquid phase of dopants and/or functional composites for different applications such as photocatalytic ones.

Also, polymer-derived ceramics (PDCs) are a route to obtain GGC in the form of powders (figure 9.12(d)), layers or fibers that are uncommonly obtained by traditional powder techniques. The PDC route is based in obtaining a pre-ceramic with tailored shape from some polymers such as polysiloxanes, polysilazanes, or polycarbosilanes, etc, using conventional polymer forming techniques. After that, the thermal decomposition of PDC in liquid, melt or organic phase is done, which produces materials with high chemical resistance, good thermomechanical properties like high creep resistance and low sintering temperatures [68].

The last step for the synthesis of GGCC is the sintering process, but traditional techniques for this purpose like pressureless sintering usually require high temperatures (1300 °C–1500 °C) and long periods of time to obtain dense materials [70]. However, for using graphene-related materials as fillers in the glass-ceramic matrix, the modification of the sintering temperature is necessary to avoid graphene degradation. Thus, novel sintering processes have been developed with shorter dwell times and low-to-mild temperature ranges. Some examples are hot pressing (HP) and hot isostatic pressing (HIP), which can sinter GGCC at low temperatures by the application of pressure, whereas the spark plasma sintering (SPS) and microwave sintering (MS) techniques are based in the use of plasma and electromagnetic fields, respectively, for obtaining high thermal loads in short times [71, 72].

9.4 Summary and outlook

To deal with the water pollution problems in the 21st century, progress in photocatalysis is of great interest. The methods to improve photocatalytic materials by different routes such as doping, heterojunction and band structure engineering. are not enough for sustainable remediation of wastewater outside laboratory. The incorporation of the photocatalysts into a supporting matrix leads to the possibility of scaling up water remediation to industrial scale. Consequently, graphene glass-ceramic composites and hybrids are attractive alternatives for the fabrication of more efficient, lighter and compact supported photocatalysts with good chemical, thermal and mechanical stability, specific structure, and tailored characteristics to use in artificial facilities or in natural water bodies at lower cost.

However, technological issues such as graphene homogenous incorporation to the vitreous matrix, or the processing temperatures, limit the short-term commercialization and large-scale applicability of these new P-GGCCs. More investigation on the integration of graphene/photocatalysts in the glass-ceramic matrix as well as reduced temperature annealing processes are needed to overcome the problems in the standardization of synthesis routes, control of graphene agglomeration, components homogenization and the whole-process reproducibility.

Acknowledgments

This work was financed through the CONACYT 40798 grant and SIP-IPN 2024–0907 project. MOR and RVTH are financed by CONACYT-PhD grants.

References

[1] Chen D *et al* 2020 Photocatalytic degradation of organic pollutants using TiO_2-based photocatalysts: a review *J. Clean. Prod.* **268** 121725

[2] Helmer R and Hespanhol I 2017 *Water Pollution Control: a Guide to the Use of Water Quality Management Principles* (CRC Press)

[3] Sibhatu A K, Weldegebrieal G K, Sagadevan S, Tran N N and Hessel V 2022 Photocatalytic activity of CuO nanoparticles for organic and inorganic pollutants removal in wastewater remediation *Chemosphere* **300** 134623

[4] Qadir M *et al* 2020 Global and regional potential of wastewater as a water, nutrient and energy source *Nat. Resour. Forum* **44** 40–51

[5] Brown J A 2005 The early history of wastewater treatment and disinfection *Proc. 2005 World Water Environmental Resources Congress* 1–7

[6] Sonune A and Ghate R 2004 Developments in wastewater treatment methods *Desalination* **167** 55–63

[7] Singh G, Singh A, Singh P, Gupta A, Shukla R and Mishra V K 2021 Sources, fate, and impact of pharmaceutical and personal care products in the environment and their different treatment technologies *Microbe Mediated Remediation of Environmental Contaminants.* (Woodhead) pp 391–407

[8] Dos Anjos N D F R 1998 Source book of alternative technologies for freshwater augmentation in Latin America and the Caribbean *Int. J. Water Resour. Dev.* **14** 365–98

[9] Kumar P S, Saravanan A, Senthil Kumar P and Saravanan Á A 2018 Sustainable waste water treatment technologies *Detox Fashion* (Springer) pp 1–25

[10] Forgacs E, Cserháti T and Oros G 2004 Removal of synthetic dyes from wastewaters: a review *Environ. Int.* **30** 953–71

[11] Crini G and Lichtfouse E 2019 Advantages and disadvantages of techniques used for wastewater treatment *Environ. Chem. Lett.* **17** 145–55

[12] López Ramírez M Á *et al* 2021 Advanced oxidation as an alternative treatment for wastewater. a review *Enfoque UTE* **12** 76–87

[13] Castaño L I, Herrera G M D and Castañeda D S G 2020 Tratamiento de aguas residuales por fotocatálisis heterogénea: una revisión sistemática *Rev. Fac. Ciencias Básicas* **16** 51–64

[14] Deng Y and Zhao R 2015 Advanced oxidation processes (AOPs) in wastewater treatment *Curr. Pollut. Reports* **1** 167–76

[15] Kumar V and Shah M P 2021 Advanced oxidation processes for complex wastewater treatment *Advanced Oxidation Processes for Effluent Treatment Plants* (Elsevier) pp 1–31

[16] Comninellis C, Kapalka A, Malato S, Parsons S A, Poulios I and Mantzavinos D 2008 Advanced oxidation processes for water treatment: advances and trends for R&D *J. Chem. Technol. Biotechnol.* **83** 769–76

[17] Palit S and Hussain C M 2021 Advanced oxidation processes as nonconventional environmental engineering techniques for water treatment and groundwater remediation *Handbook of Advanced Approaches Towards Pollution Prevention and Control* **1** (Elsevier) pp 33–44

[18] Kamali M, Aminabhavi T M, Maria M E, Ul Islam S, Appels L and Dewil R 2023 Homogeneous advanced oxidation processes for the removal of pharmaceutically active compounds—current status and research gaps *Advanced Wastewater Treatment Technologies for the Removal of Pharmaceutically Active Compounds* (Springer) pp 181–210

[19] Bhatti D T and Parikh S P 2022 Recent progress in doped TiO_2 photocatalysis and hybrid advanced oxidation processes for organic pollutant removalfrom wastewater *Curr. World Environ.* **17** 146–60

[20] Alabdraba D W M S, Al-Obaidi A R, Hashim S S and Zangana S D 2018 Industrial wastewater treatment by advanced oxidation processes—a review *J. Adv. Sci. Eng. Technol.* **1** 24–33

[21] Poyatos J M, Muñio M M, Almecija M C, Torres J C, Hontoria E and Osorio F 2010 Advanced oxidation processes for wastewater treatment: state of the art *Water Air Soil Pollut.* **205** 187–204

[22] Arslan-Alaton I 2003 A review of the effects of dye-assisting chemicals on advanced oxidation of reactive dyes in wastewater *Color. Technol.* **119** 345–53

[23] Garcés Giraldo C, Fernando L, Franco M, Alejandro E, Arango S and Julián J 2004 La fotocatálisis como alternativa para el tratamiento de aguas residuales *Rev. Lasallista Investig.* **1** 83–92 https://redalyc.org/articulo.oa?id=69511013 (accessed 22 November 2023)

[24] Amakiri K T, Angelis-Dimakis A and Canon A R 2022 Recent advances, influencing factors, and future research prospects using photocatalytic process for produced water treatment *Water Sci. Technol.* **85** 769–88

[25] Ali H M, Arabpour Roghabadi F and Ahmadi V 2023 Solid-supported photocatalysts for wastewater treatment: supports contribution in the photocatalysis process *Sol. Energy* **255** 99–125

[26] Chouhan A P S and Sarma A K 2011 Modern heterogeneous catalysts for biodiesel production: a comprehensive review *Renew. Sustain. Energy Rev.* **15** 4378–99

[27] Akçağlar S 2022 Preparation of MoO3/MoS2-E composite for enhanced photoelectrocatalytic removal of antimony from petrochemical wastewaters *Turkish J. Chem.* **46** 1450–67

[28] Borges M E, de Paz Carmona H, Gutiérrez M and Esparza P 2023 Photocatalytic removal of water emerging pollutants in an optimized packed bed photoreactor using solar light *Catalysts* **13** 1023

[29] Abu Bakar Sidik D *et al* 2018 Photocatalytic degradation of industrial dye wastewater using zinc oxide-polyvinylpyrrolidone nanoparticles *Malaysian J. Anal. Sci.* **22** 693–701

[30] Ganapathy N R V, Krishnan V, Veeraraghavan A J, V and Devaraj G 2019 Biosynthesis of zinc oxide nanoparticles from plectranthus amboinicus and its photocatalytic effect on wastewater treatment *Int. J. Recent Technol. Eng.* **8** 660–3

[31] Aremu O H, Akintayo C O, Naidoo E B, Nelana S M and Ayanda O S 2021 Synthesis and applications of nano-sized zinc oxide in wastewater treatment: a review *Int. J. Environ. Sci. Technol.* **18** 3237–56

[32] Mashuri S I S *et al* 2020 Photocatalysis for organic wastewater treatment: from the basis to current challenges for society *Catalysts* **10** 1260

[33] Okpara E C, Olatunde O C, Wojuola O B and Onwudiwe D C 2023 Applications of transition metal oxides and chalcogenides and their composites in water treatment: a review *Environ. Adv.* **11** 100341

[34] Pikula K *et al* 2020 Aquatic toxicity and mode of action of CdS and ZnS nanoparticles in four microalgae species *Environ. Res.* **186** 109513

[35] Xiao Z *et al* 2020 Materials development and potential applications of transparent ceramics: a review *Mater. Sci. Eng. R: Rep.* **139** 100518

[36] Ramírez C, Belmonte M, Miranzo P and Osendi M I 2021 Applications of ceramic/graphene composites and hybrids *Mater* **14** 2071

[37] Sakamoto A and Yamamoto S 2010 Glass–ceramics: engineering principles and applications *Int. J. Appl. Glas. Sci.* **1** 237–47

[38] Casasola R, Rincón J M and Romero M 2011 Glass–ceramic glazes for ceramic tiles: a review *J. Mater. Sci.* **47** 553–82

[39] Wang J, Wang M, Tian Y and Deng W 2022 A review on photocatalytic glass ceramics: fundamentals, preparation, performance enhancement and future development *Catalysts* **12** 1235

[40] Singh G, Sharma M and Vaish R 2021 Emerging trends in glass-ceramic photocatalysts *Chem. Eng. J.* **407** 126971

[41] Höland W, Rheinberger V and Schweiger M 2003 Control of nucleation in glass ceramics *Philos. Trans. R. Soc. London. Ser. A Math. Phys. Eng. Sci.* **361** 575–89

[42] Yazawa T, Machida F, Oki K, Mineshige A and Kobune M 2009 Novel porous TiO_2 glass-ceramics with highly photocatalytic ability *Ceram. Int.* **35** 1693–7

[43] Abdel-Wahed M S, Abdel-Karim A, Margha F H and Gad-Allah T A 2021 UV sensitive ZnO and TiO_2-ZnO nanocrystalline transparent glass-ceramic materials for photocatalytic decontamination of surface water and textile industry wastewater *Environ. Prog. Sustain. Energy* **40** e13653

[44] Gaur A, Sharma M, Chauhan V S and Vaish R 2023 $BaTiO_3$ crystallized glass-ceramic for water cleaning application via piezocatalysis *Nano-Struct. Nano-Objects* **35** 101005

[45] Zhang J *et al* 2022 Recent progress on carbon-nanotube-based materials for photocatalytic applications: a review *Sol. RRL* **6** 2200243

[46] Li X, Yu J, Wageh S, Al-Ghamdi A A and Xie J 2016 Graphene in photocatalysis: a review *Small* **12** 6640–96

[47] Xie K, Fang J, Li L, Deng J and Chen F 2022 Progress of graphite carbon nitride with different dimensions in the photocatalytic degradation of dyes: a review *J. Alloys Compd.* **901** 163589

[48] Abuzeyad O H, El-Khawaga A M, Tantawy H and Elsayed M A 2023 An evaluation of the improved catalytic performance of rGO/GO-hybrid-nanomaterials in photocatalytic degradation and antibacterial activity processes for wastewater treatment: a review *J. Mol. Struct.* **1288** 135787

[49] Gnanaseelan N, Latha M, Mantilla A, Sathish-Kumar K and Caballero-Briones F 2020 The role of redox states and junctions in photocatalytic hydrogen generation of MoS_2-TiO_2-rGO and CeO_2-$Ce_2Ti_3O_{8.7}$-TiO_2-rGO composites *Mater. Sci. Semicond. Process.* **118** 105185

[50] Gnanaseelan N, Marasamy L, Mantilla A, Kamaraj S K, Espinosa-Faller F J and Caballero-Briones F 2022 Cesium-decorated reduced graphene oxide for photocatalytic hydrogen generation *Mater. Lett.* **314** 131864

[51] Xu T, Zhang L, Cheng H and Zhu Y 2011 Significantly enhanced photocatalytic performance of ZnO via graphene hybridization and the mechanism study *Appl. Catal. B Environ.* **101** 382–7
[52] Zhu M, Chen P and Liu M 2011 Graphene oxide enwrapped Ag/AgX (X = Br, Cl) nanocomposite as a highly efficient visible-light plasmonic photocatalyst *ACS Nano* **5** 4529–36
[53] Fu Y and Wang X 2011 Magnetically separable $ZnFe_2O_4$-graphene catalyst and its high photocatalytic performance under visible light irradiation *Ind. Eng. Chem. Res.* **50** 7210–8
[54] Ikram M, Raza A, Imran M, Ul-Hamid A, Shahbaz A and Ali S 2020 Hydrothermal synthesis of silver decorated reduced graphene oxide (rGO) nanoflakes with effective photocatalytic activity for wastewater treatment *Nanoscale Res. Lett.* **15** 1–11
[55] Porwal H *et al* 2013 Toughened and machinable glass matrix composites reinforced with graphene and graphene-oxide nano platelets *Sci. Technol. Adv. Mater.* **14** 055007
[56] Porwal H *et al* 2014 Tribological properties of silica–graphene nano-platelet composites *Ceram. Int.* **40** 12067–74
[57] Huang Y, Yasuda K and Wan C 2020 Intercalation: constructing nanolaminated reduced graphene oxide/silica ceramics for lightweight and mechanically reliable electromagnetic interference shielding applications *ACS Appl. Mater. Interfaces* **12** 55148–56
[58] Porwal H, Grasso S, Cordero-Arias L, Li C, Boccaccini A R and Reece M J 2014 Processing and bioactivity of 45S5 Bioglass®-graphene nanoplatelets composites *J. Mater. Sci. Mater. Med.* **25** 1403–13
[59] Porwal H, Grasso S and Reece M J 2013 Review of graphene–ceramic matrix composites *Adv. Appl. Ceram.* **112** 443–54
[60] Sharma P, Sharma G and Punia R 2022 Graphene: a prime choice for ceramic composites *Advanced Ceramics for Versatile Interdisciplinary Applications* (Elsevier) pp 417–35
[61] David L, Bhandavat R, Barrera U and Singh G 2016 Silicon oxycarbide glass-graphene composite paper electrode for long-cycle lithium-ion batteries *Nat. Commun.* **7** 1–10
[62] Amiri A, Naraghi M, Ahmadi G, Soleymaniha M and Shanbedi M 2018 A review on liquid-phase exfoliation for scalable production of pure graphene, wrinkled, crumpled and functionalized graphene and challenges *FlatChem.* **8** 40–71
[63] Niu L, Coleman J N, Zhang H, Shin H, Chhowalla M and Zheng Z 2016 Production of two-dimensional nanomaterials via liquid-based direct exfoliation *Small* **12** 272–93
[64] Sharma N, Saxena T, Alam S N, Ray B C, Biswas K and Jha S K 2022 Ceramic-based nanocomposites: a perspective from carbonaceous nanofillers *Mater. Today Commun.* **31** 103764
[65] Miranzo P, Belmonte M and Osendi M I 2017 From bulk to cellular structures: a review on ceramic/graphene filler composites *J. Eur. Ceram. Soc.* **37** 3649–72
[66] Ilyas K *et al* 2019 In-vitro investigation of graphene oxide reinforced bioactive glass ceramics composites *J. Non. Cryst. Solids* **505** 122–30
[67] Yilmaz E and Soylak M 2020 Functionalized nanomaterials for sample preparation methods *Handbook of Nanomaterials in Analytical Chemistry: Modern Trends in Analysis* (Elsevier) pp 375–413
[68] Song C, Liu Y, Ye F, Wang J and Cheng L 2020 Microstructure and electromagnetic wave absorption property of reduced graphene oxide-SiCnw/SiBCN composite ceramics *Ceram. Int.* **46** 7719–32
[69] Elsayed H, Picicco M, Dasan A, Kraxner J, Galusek D and Bernardo E 2020 Glass powders and reactive silicone binder: application to digital light processing of bioactive glass-ceramic scaffolds *Ceram. Int.* **46** 25299–305

[70] Liu H, Lu H, Chen D, Wang H, Xu H and Zhang R 2009 Preparation and properties of glass–ceramics derived from blast-furnace slag by a ceramic-sintering process *Ceram. Int.* **35** 3181–4

[71] Gao C, Feng P, Peng S and Shuai C 2017 Carbon nanotube, graphene and boron nitride nanotube reinforced bioactive ceramics for bone repair *Acta Biomater.* **61** 1–20

[72] Mandal A K and Sen R 2017 An overview on microwave processing of material: a special emphasis on glass melting *Mater. Manuf. Process.* **32** 1–20

IOP Publishing

Glass-based Materials
Advances in energy, environment and health
Sathish-Kumar Kamaraj and Arun Thirumurugan

Chapter 10

Recycled glass used in mortars and reinforced concrete for tropical marine environments

Edwin Hoil-Canul, Khirbet López-Velázquez, José Luis Cabellos, Juan A Ríos-González, L Maldonado-López and L Díaz-Ballote

Cement has a seminal role for binding concrete aggregates and sand in mortars. Its production has been constantly growing. In 1995, global production was just 1.39 billion tons and in 2022 reached 4.1, almost the production forecasted for the year 2050. Such an increment demands huge amounts of resources and energy and may also contribute to the generation of greenhouse gasses. However, the use of recycled glass powder (RGP) may benefit the environment, saving energy and raw resources. Recycled glass is increasingly suggested as a coarse and fine aggregate but also fine ground as a cementitious material because it reacts with other cement components, acting as a pozzolanic material and consequently diminishes the porosity of concrete. This property makes it adequate for reinforced concrete exposed to very harsh environments, where rebars corrosion is catalyzed by chlorides in the marine breeze and high temperatures and relative humidity values typical for tropical environments. In this chapter, a review of the use of different types of recycled glass in mortars and reinforced concrete and the corrosion behavior of glass concrete exposed to the marine breeze, are presented.

10.1 Introduction

Atmospheric marine environments can accelerate the corrosion processes, mainly due to the effects of chloride ions. The ingress of those ions in concrete has been studied by experimental and computational schemes. The chloride aggressiveness depends on the wind, decreasing within a few hundred meters from the sea [1]. Chloride ions may penetrate the concrete through capillary absorption, hydrostatic pressure, or diffusion. For durable concrete aimed for marine environments, permeability through the pores is the most important characteristic to avoid chloride ions penetration [2]. The use of RGP has been evaluated as a partial cement

doi:10.1088/978-0-7503-5904-7ch10

replacement in concrete, reporting that the glass addition can reduce the cost of concrete use up to 14% [3]. In addition, the production of six tons of glass powder concrete can reduce one ton of CO_2 emissions, contributing significantly to environment protection by reducing greenhouse gases emission and particulate production [3, 4]. Furthermore, the increase of the workability of concrete mixed with glass powder has been observed due to the lower water affinity, smoother surface, and the lower fresh density, which was attributed to the low specific gravity of glass powder. On the other hand, the addition of glass powder in concrete as cement replacement can improve hardened density with curing time and contributes to the resistance of corrosion. Tamanna and Tuladhar [5] reported that the replacement of cement with glass powder did not show significant strengthening at an early age due to the pozzolanic reaction; however, a notable strength development was observed until 56 days. Moreover, in agreement with Du and Tan [6], the optimum glass powder content in concretes ought to be between 10% and 20%. Therefore, RGP can be considered as a supplementary cementitious material for sustainable concrete practice. However, it is important to notice that most data were obtained under controlled conditions and the corrosion behavior of rebar in concrete with RGP remains almost unexplored. Although some authors have investigated such behavior under controlled conditions, others have reported some results after two years of exposure in field conditions [4, 8].

Recycling of materials is a seminal subject, mainly when the production of such materials from raw materials requires an important energy amount and contributes to the emission of greenhouse gasses. As a notable example, we have Portland cement production, which has been increasing constantly due to a growing world population which requires housing, transportation, hospitals, and buildings for a comfortable life. Global cement production was 2.54 Gt in 2006, and it has been estimated to increase up to 18 Gt by 2050 [9–11]. For this reason, a great effort has been made to reduce the energy consumption in obtaining clinker for cement production, by means of different methods and production procedures. Also, recycled materials, new binders, biopolymers, etc. Glass is used in daily life, including disposable beverage bottles and jars of single-use, whose final disposal represents major challenges for the municipal waste collection systems and landfills, since glass takes thousands of years to decompose. Glass production is increasing constantly. In the UK, the Department of Environment, Food, and Rural Affairs reported that the UK generated 2.4 million tons of waste glass in 2017, claiming that only 67.6% of the post-consume glass produced in the UK was collected for recycling [12]. However, the EU aims to recycle 90% of glass containers and jars by 2030 [13], Sweden being a model country for glass collection and recycling, since they recycle 95% of glass residues [14]. In Mexico, recycled glass is processed mainly for the manufacturing of new bottles; however, only 12% of the waste glass is recycled in the country [15], due to the fact that coloured glass is not accepted for recycling and bad collection strategies. Recently, recycled glass has been proposed for infraectructure construction worldwide. In Mexico, there is not enough research aiming at the evaluation of the properties and behavior of glass concrete prepared with different aggregate and exposed to diverse industrial or natural microclimates such as marine environments. Properties such as fresh density, air content, slump, compressive strength, splitting tensile strength, flexural strength, modulus of elasticity, drying shrinkage, ASR (alkali–silica reaction), among others must be widely studied (table 10.1). The findings presented in this work attempt to contribute to the investigations of some of these properties.

Table 10.1. Main properties and ASTM standard for concrete and mortars testing.

Property	Number of specimens required	Test age (days)	Dimensions (mm)	Standard
Fresh density	—	—	—	ASTM C138 EN 12350-6
Air content	—	—	—	ASTM C138 EN 12350-7
Slump	—	—	—	ASTM C143 AS 1012.3.1 EN 12350-2
Compressive strength	9	7, 28, 90	100 × 100 × 100	ASTM C109 BS N 12390-3 AS 1012.9 EN 196-1
Flexural strength	3	28	100 × 100 × 400	ASTM C78 AS 1012.11 EN 196-1
Elastic modulus	3	28	Ø100 × 200 (cylinder)	ASTM C215, C469 EN ISO 9856
Splitting tensile strength	3	28	Ø100 × 200 (cylinder)	ASTM C496 AS 1012.10 EN 196-1
Drying shrinkage	2	up to 56	75 × 75 × 285	ASTM C157 EN 680
Rapid chloride permeability test (RCPT)	6	28, 90	Ø100 × 50	ASTM C1202

(*Continued*)

Table 10.1. (*Continued*)

Property	Number of specimens required	Test age (days)	Dimensions (mm)	Standard
ASR	3	up to 49	25 × 25 × 285	ASTM C1260
Absorbability	6	28	40 × 40 × 160	ASTM C1585 PN-EN 1097-6: 2011 UNE 83980

10.2 Recycled glass in mortar and concrete

10.2.1 Mortars

Mortars are a mixture of cement, fine aggregate, and water, in a ratio of one part of cement to 2.25–3.5 of fine aggregate [16]. They are mainly used for masonry applications, decorative purposes or to avoid environmental pollution. In recent decades, several research groups have focused on analyzing the possibilities of using crushed recycled glass as a fine and coarse aggregate or fine ground glass as a replacement for cement in concrete and mortars. There is an agreement on the idea that the recycling of glass waste is a promising pro-ecological measure, since its use in concrete provides a potential solution for the decline of natural sand sources worldwide [17]. Replacing 10% of cement in standard mortars with glass powder resulted in a higher resistance to sulfates without reducing strength [18, 19]. Huseien *et al* [20], investigated the effects of recycled glass ground as nanopowder in mortars (proportion of 5%), with granulated blast furnace slag, observing an increase of 40% in the tensile strength. Long *et al* [21], studied the encapsulation of Pb leaching, from waste cathode ray tube (CRT) glass, through fly ash-slag geopolymer mortars, finding that the recycled CRT glass enhanced both the physical encapsulation and the chemical solidification of Pb particles, as the silica modulus increased; the compressive strength and the ASR expansion first increased and then decreased, and the leached concentration of Pb decreased significantly. The increased silica modulus improved the chemical binding of Pb ions by generating lead silicate. When mechanical strength, ASR expansion and Pb leaching rate were considered, the authors concluded that alkali dosage of 6% and a silica modulus of 1.5 are optimal conditions for recycling CRT glass in geopolymer mortars. Sikora *et al* [22] used recycled brown glass in ratios of 25%, 50%, 75% and 100% as replacement for quartz sand, cement containing TiO_2 and nanosilica particles to prepare self-cleaning and bactericidal cement mortar composites. They reported that recycled waste glass may result in a successful replacement of sand in cement mortars when nanosilica particles are incorporated. In addition, properties such as the flexural strength diminished with the glass content but reached the same value as the control sample with 100% glass replacement when nanosilica particles were added. They also found that the compressive strength was higher for samples containing recycled glass and nanosilica than for the reference sample. Furthermore, it was shown that the bacteria, *Escherichia coli*, was inactivated in about 30 min, when it was deposited on the cement mortar surface, mainly due to a higher porosity exhibited by the mortars surface, which allowed the TiO_2 particles to act more efficiently. Zhang *et al* [23] have studied the influence of the particle size on mortars and their mechanical properties with recycled glass proportion of 5%, 10%, 15%, and 20%. Glass powders were obtained by different grinding times resulting in average particle sizes of 1670.0, 243.0, and 13.2 μm. The particles of 13.2 μm had a specific surface area of 670 m^2 kg^{-1} and high pozzolanic activity. It is worth noting that several authors corroborate that the pozzolanic activity is strongly influenced by the glass particle size [24–26].

10.2.2 Concrete

Concrete is designed based on the application. It is essentially a mixture of cement, fine and coarse aggregates, and water in different proportions. Cement reacts with water and sets resulting in properties such as flexural and compressive mechanical strength, workability, water absorption, etc (table 10.1). Due to its nature, concrete is a very porous material influenced by several factors such as coarse and fine aggregate types, mixture design, cement type, etc. Concrete and mortars are the most used construction materials worldwide and as the world population grows their demand increases. As a result, huge amounts of raw resources are needed, threatening the production sustainability. Additionally, the cement industry contributes to environmental greenhouse gas emissions by preparing concrete and mortars [27]. Therefore, alternative materials, mainly recycled by-products, are being proposed to replace either totally or partially the classical cement and natural raw material in mortars and concrete [7, 23]. By these means, recycled glass and materials play a key role in regions where industrial by-products are scarce, or not available. On the other hand, beverage glass containers and jars from diverse devices are present almost worldwide. Fine ground glass may be utilized as a partial replacement to produce Portland concrete mixtures due to its pozzolanic properties [28]. Moreover, the inclusion of recycled glass in proportion up to 20%–30% improves concrete strengthening significantly [12]. In addition, it was reported that by replacing natural sand to 20% of glass from bottles, the mechanical properties were enhanced, 20% being the optimal glass content to achieve an excellent workability [27]. Properties of fresh concrete may be improved by using glass as a fine aggregate due to the smooth surface and relatively low water absorption [29]. Although several authors have studied different ratios of recycled waste glasses as aggregates for concrete and mortars [6, 30], it has been noticed that a mass proportion beyond 20%–30% of glass against conventional aggregates has not displayed any significant influence in the properties of concrete [6, 17, 31]. Fresh and hardened properties of concrete and mortars with recycled glass as aggregate have been extensively investigated. Durability tests, including ASR were also evaluated and no negative effects were observed [32]. In fact, compared to natural sand, glass sand has a higher fraction of microparticles, and its pozzolanic properties are influenced by its particle size distribution. Furthermore, it has been observed that glass powder with particle size lower than 300 μm reduces the ASR expansion and micro-cracks in the concrete [29]. Fine aggregates may come from crushed rocks or natural river sand graded based on a specific modulus fineness. Recently, crushed and ground particles of recycled glass from different sources have been suggested as sand replacement. Chen *et al* [33] proposed the replacement of natural river sand by ground colorless and transparent flat glass from waste using different proportions (0%, 20%, 40%, 60% and 80%). It was found that the optimal replacement was 20%, since such proportion promoted a better cement hydration compared to the others. Also, the porosity and the gas permeability decreased to a 16.5% and 57.4%, respectively. Furthermore, compressive strength increased by 3% and the elastic modulus was higher by 5.9%. Shaker *et al* [27], studied the use of recycled glass waste in concrete as a partial or complete replacement for aggregates, noticing that several authors reported an increment of the workability because of a weaker cohesion between the cement mortar and the smooth surface of the glass particles. In contrast to the increase in the workability, when the glass particle edges were sharp and had angular forms, the movement amongst the cement mortar and the particles was blocked. Most studies have reported a decrease in the mechanical strength, associated mainly to a smoother surface and the sharp edges of the waste glass, which cause a poor adhesion to the cement mortar and glass particles and an increment in water content due to the poor water absorption of the glass. Moreover, in some cases, the ASR can result in cracks.

10.3 Concrete with recycled glass powder exposed to marine conditions

Concrete from ordinary Portland cement provides a protective cover to the embedded rebars. Additionally, it assures an adequate pH environment of about 12.5 in the pore solution, allowing the chemical stability of the rebars, better known as passivity. However, pollutants in the atmosphere, such as Cl^-, CO_2, SO_2, and humidity among others, may penetrate through the concrete pores and voids and cause the concrete to deteriorate by carbonation and corrosion of the rebars, affecting the durability and integrity of the infrastructure. Under marine environments such effects may be catastrophic.

10.3.1 Carbonation of the concrete

Carbonation of concrete proceeds mainly in two steps [34]. First, the water reacts with carbon dioxide from the atmosphere to build up carbonic acid (equation (10.1)). Then, the calcium hydroxide in the solution pore and the carbonic acid form calcium carbonate (equation (10.2)), or the carbonic acid and the calcium silicate hydrates form calcium carbonate, silicon oxides, and water (equation (10.3)).

$$CO_2 + H_2O \rightarrow H_2CO_3 \quad (10.1)$$

$$Ca(OH)_2 + H_2CO_3 \rightarrow CaCO_2 + 2H_2O \quad (10.2)$$

$$C_3S_2H_3 + 3H_2CO_3 \rightarrow 3CaCO_2 + 2SiO_2 + 6H_2O \quad (10.3)$$

During the reaction phases, the pH of the pore solution diminishes. There are qualitative and quantitative methods to measure the degree of carbonation, i.e., the decrease of pH [35, 36]. In this study, a qualitative method based on the phenolphthalein indicator, was chosen to measure the carbonation depth of the concrete exposed to the marine breeze.

10.3.2 Reinforcing bar corrosion

The onset of corrosion starts when chloride ions from the air or in the marine spray penetrate the pores in the concrete and reach the surface of the embedded rebars. It begins as pitting corrosion on the passive film, but if a chloride threshold is fulfilled, corrosion takes place. Corrosion current density is used as a criterion to evaluate the state of corrosion of the rebars. While 0.1 $\mu A\ cm^{-2}$ is considered indicative of the onset of corrosion, values among 0.5 and 1 $\mu A\ cm^{-2}$ are considered a state of moderate-to-high corrosion; thereafter, it is considered that the rebars are under high corrosion state [37, 38] (check figure 10.5).

There is not a specific value for the chloride threshold to initiate corrosion on the bars, since it depends on the concrete mixture design, the materials used to prepare the concrete, the rebars surface conditions, etc. Nonetheless, values among 0.40 ppcw (percentage per cement weight) and 1.2 ppcw [7] are considered to evaluate the degree of corrosion of the rebars, as shown in figure 10.5.

10.4 Corrosion of reinforced concrete with RGP, exposed to marine conditions

There is scarce information about the performance of reinforced concrete prepared with RGP under marine conditions. Nevertheless, Peng *et al* [4], carried out laboratory rapid tests, simulating corrosive marine environments, on reinforced recycled aggregate concrete with RGP as cement replacement using proportions of 0% (as reference) 10%, 20%, 30% and 40%. They studied the chloride penetration, among other properties, finding that the use of RGP in concrete with recycled aggregates, increased the chloride penetration and corrosion resistance with the compressive mechanical strength (CMS) remaining unaffected. They attributed such behavior to a refinement of the pore structure by the formation of secondary C–(N)–S–H gels with a lower Ca/Si ratio, and to the increase of alkalinity of pore solution due to the depolymerization of GP and the released Na+ from the RGP. Recently, Hoil *et al* [7], replaced 0% (reference), 5%, 10%, and 15% of cement for RGP coming from mixed color bottles, in concrete prepared with crushed limestone aggregates, and water-to-cement (w/c) ratios of 0.4, 0.5 and 0.7. They measured the CMS after 28, 90, and 270 days of curing time in a saturated solution of lime and water. On the other hand, samples were prepared to be exposed to the marine breeze in field. After two years of exposure to the marine breeze, they reported a beneficial corrosion effect by the addition of RGP, mainly in mixtures with higher water content, with ratios of 0.5 and of 0.7, compared to control samples. In the forthcoming sections the CMS evolution after nine months of curing and the corrosion behavior in reinforced samples exposed for 48 months will be discussed.

10.4.1 CMS after nine months

After nine months of curing, beneficial effects on the CMS were displayed in the concrete because of the glass powder content, table 10.2 and figure 10.1. This finding may be attributed to a refinement of the concrete pore microstructure due to the release of Na^+ present in the glass powder [4]. Nevertheless, it is important to mention that the increment in the CMS is not too large. For a replacement of cement for 15% of RGP and a w/c ratio of 0.4, 0.5 and 0.7, the CMS was on average 38 MPa, 44 MPa and 41 MPa, respectively. The above may be explained by the fact that by refining the pore microstructure of the concrete by polymerization of the SiO_2 present in the RGP, the pore solution should be strongly alkaline and, in this study, the cement replacement by glass powder is too low to release enough Na^+ ions to alkalize the pore solution.

The pore structure is modified by the RGP, but it seems that it is not enough to enhance the CMS and mitigate the concrete carbonation, even if the pore volume decreases when the RGP content increases, mainly for w/c ratios lower than 0.4 (check figure 10.2). Even if the standard test ASTM C642-21 [39] does not consider

Table 10.2. Average CMS (MPa) after 9 months of curing in a saturated lime solution. Δ was calculated only for 15% of the cement replaced by recycled glass powder.

	Recycled glass powder (%)				
a/c	0	5	10	15	Δ
0.4	42.0 ± 4.0	42.0 ± 0.5	42.0 ± 0.4	38.0 ± 0.2	– 4
0.5	33.0 ± 0.3	38.0 ± 1.6	43.0 ± 0.2	44.0 ± 0.1	11
0.7	31.5 ± 0.3	38.7 ± 0.8	32.4 ± 0.5	40.6 ± 0.5	9

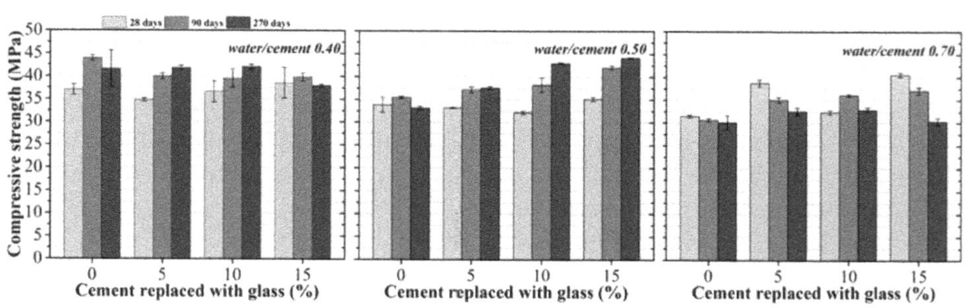

Figure 10.1. Compressive mechanical strength after nine months of curing in saturated lime solution.

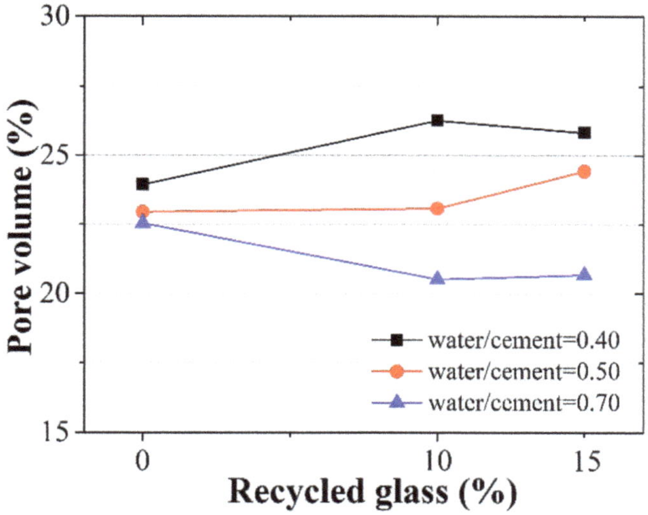

Figure 10.2. Pore volume as a function of w/c ratio and RGP content.

the total pore content in the hardened concrete because it involves an apparent density, it may present a qualitative approach to the pore content and voids in concrete, figure 10.2.

It can be observed that the pore volume percentage decreases when cement is replaced for 10% and 15% of glass powder. Replacing cement for 5% seemed not to influence the pore volume in the mixture. Air voids appear mainly in low slump concrete and may affect the CMS by increasing the levels CO_2, causing the carbonation of the concrete and also the Cl^- penetration resistance and the humidity ingress, which may cause corrosion of the rebars. Even though the CMS did not rise dramatically and did not surpass the CMS for concrete with a w/c ratio of 0.40, a beneficial effect was observed from the addition of RGP, mainly in mixtures with higher water content, as w/c of 0.50 and of 0.70, regarding the control samples (0% of cement replacement).

After 9 months of curing, the replacement of cement by glass did not have any influence on the CMS average value of the concrete with a w/c ratio of 0.40. In fact, it showed a decrease of 4 MPa with 15% of recycled glass, that may be attributed to the creation of voids in this mixture, which is too dense and hence more likely to form voids resulting in a decrease of the CMS. However, the w/c ratios of 0.50 and 0.70 showed a clear beneficial effect with the addition of glass powder due to densification of the pore structure as explained before. Additionally, the chloride ingress in these mixtures decreases strongly, as will be described in sections below.

10.4.2 Carbonation

The carbonation front was measured by taking concrete slides with radius of 37.5 mm without rebars by the phenolphtalein test [35, 36], figure 10.3. It was observed that carbonation did not affect a hypothetical rebar of 9.4 mm of radius

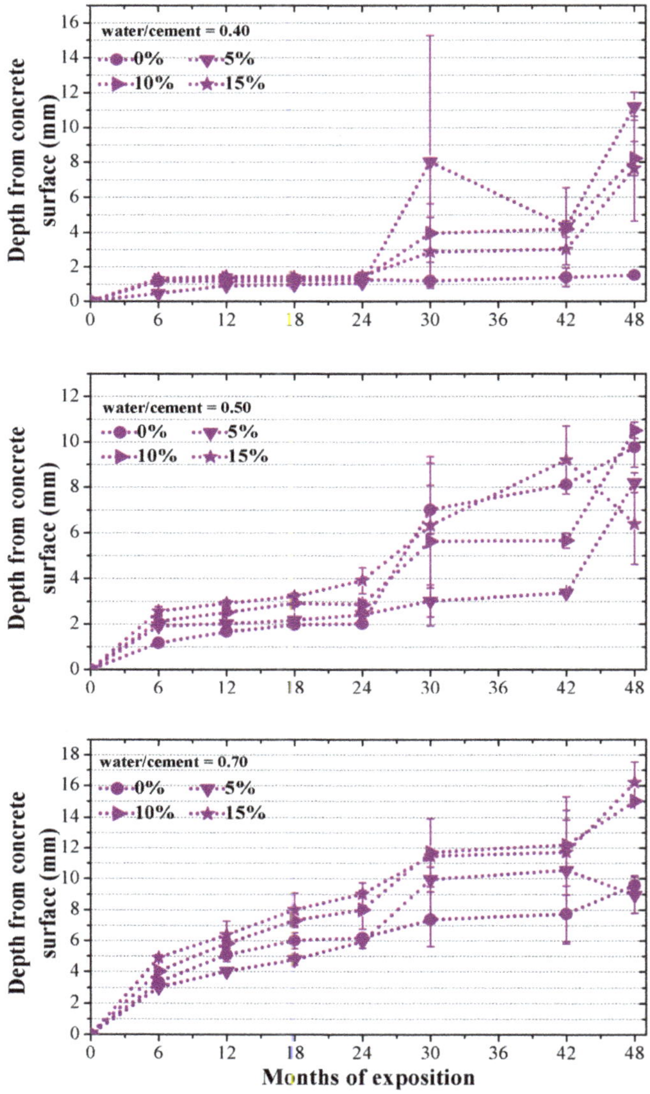

Figure 10.3. Carbonation depth in samples after 48 months of exposure to the marine breeze.

embedded in the center of the cylinders because they were located 32.5 mm far away from the surface of the samples. However, it seems that the replacement of 5%, 10% and 15% of cement for recycled glass is not enough to prevent the penetration of CO_2 from the environment, even if some densification of the pore structure was observed, figure 10.2.

Carbonation may cause corrosion because it acidifies the pore solution in the concrete and may destroy the passive film when the pore solution reaches a pH value of 9.5 [40]. Pore solution of fresh concrete exhibits a pH value of between 12.5 and 13 [41], which is a requirement to have an intact passive film on the surface of the rebars and avoid further corrosion. Because the carbonation front is no affecting the center of concrete samples, it may be assumed that the corrosion only due to chloride penetration.

10.4.3 Chloride penetration

To measure the free chloride content, powder was obtained from the rebar position by drilling concrete slides. The powder was milled until it could pass through a number 50 sieve. The water-soluble chloride was determined by volumetric titration, according to the standard practice ASTM D1411-09 [42]. The extraction process of the concrete powder to determine the chloride content is explained in detail elsewhere [7, 43].

Chloride content at the bar position showed values in the upper limit suggested in previous reports, mainly for the control samples with w/c ratio of 0.50 and 0.70 (0% of glass powder). However, it is worth mentioning that RGP decreases the chloride penetration by 90%, 60%, and 23% by replacing 5%, 10%, and 15% of cement from RGP in mixtures with a w/c ratio of 0.4. The decrease of chloride penetration was 54%, 42%, and 15% when the w/c ratio was of 0.50 and 0.70, and with similar replacement amounts of RGP, figure 10.4 and table 10.3.

There is scarce information about the behavior of reinforced concrete under marine conditions to compare the obtained results with similar conditions. However, the decrease of chloride penetration may be attributed to refinement of the concrete pore microstructure due to the release of the Na^+ present in the glass powder. Even though the increment in CMS was not significant. Overall, for a w/c ratio of 0.40, 0.50, and 0.70, after nine months of curing, the chloride ions penetration may be explained by the fact that by refining the pore microstructure of the concrete, the pore solution should be strongly alkaline and, in this study, the cement replacement by glass powder quantity is too low to release enough Na^+ ions to alkalize the pore solution [4]. In a similar set of concrete, exposed for two years to a natural tropical marine environment, Hoil *et al* [7] observed a decrease in the carbonation front and a lower chloride penetration compared to control samples. Additionally, in the reinforced concrete, the corrosion onset was delayed. The samples were exposed for two years to a very corrosive marine condition, Category C5, which is very high, according to ISO 9223: 2012 [44]. After 48 months of exposure, the carbonation penetration was not stopped by the densification of the pore structure. Some authors proposed a pore classification based on their diameters: (1) gel pores (diameter <10 nm); (2) transitional pores (diameter, 10–100 nm); (3) capillary pores (diameter 100–1000 nm) and (4) macropores (diameter $>$ 1000 nm) [45].

The concrete carbonation happens when atmospheric CO_2 is absorbed through half-empty pores solution and reacts with water to produce carbonation, according with equations (10.1)–(10.3) in section 10.4. However, if the pore is water-full the CO_2 molecules do not penetrate easily due to the size of their radius, which is 0.24 nm. Instead, the radius of the Cl^- ions is 0.181 nm [46] and they move easily in the pore structure, causing corrosion of the rebars.

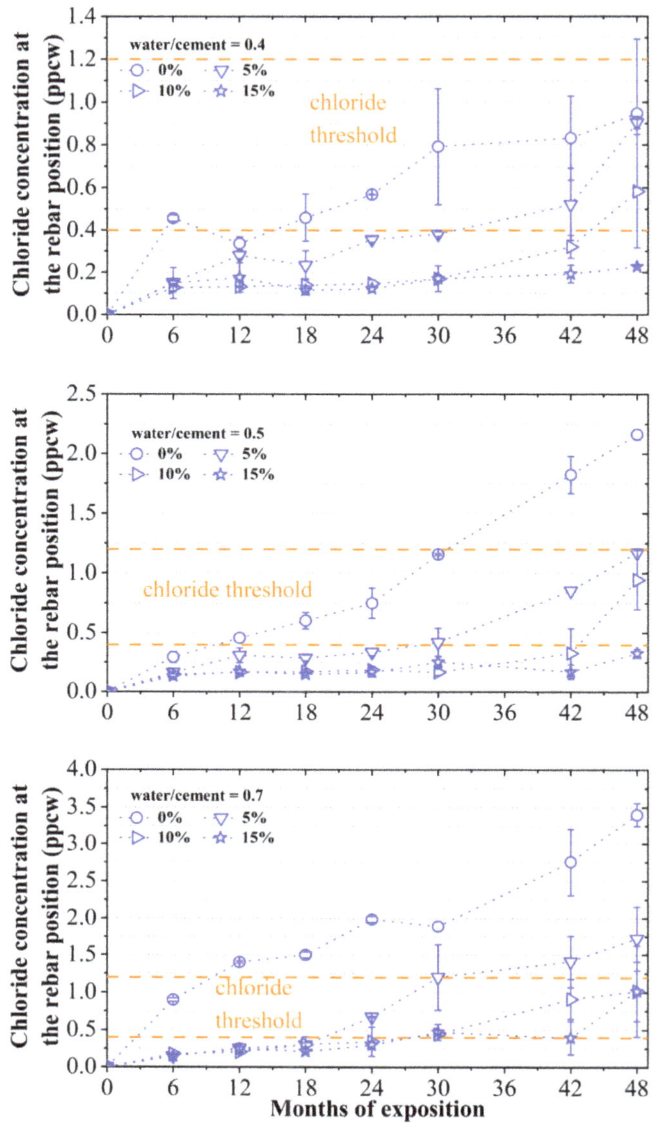

Figure 10.4. Chloride penetration at the bars position after 48 months of exposure to the marine breeze.

10.4.4 Corrosion current density

The corrosion current density showed that control rebars in concrete of a w/c ratio of 0.40, displayed a state of moderate-to-high corrosion after about 28 months of exposure, while samples with 5% of glass powder were in the region of moderate corrosion and those with 10% and 15% of RGP remained in the passive region, figure 10.5.

It is important to mention that corrosion in control samples showed an expected behavior as a direct function of the w/c ratio i.e. less corrosion current density for

Table 10.3. Chloride concentration (percentage per cement weight, ppcw) at bar position (30 mm) as a function of the cement replaced from RGP for different w/c ratios and after 48 months of exposure to the marine breeze.

	Replaced cement by recycled glass powder			
w/c	0%	5%	10%	15%
0.40	1.0 ± 0.40	0.90 ± 0.02	0.6 ± 0.03	0.23 ± 0.03
0.50	2.16 ± 0.04	1.17 ± 0.00	0.9 ± 0.25	0.33 ± 0.40
0.70	3.40 ± 0.15	1.17 ± 0.45	1.02 ± 0.30	1.02 ± 0.61

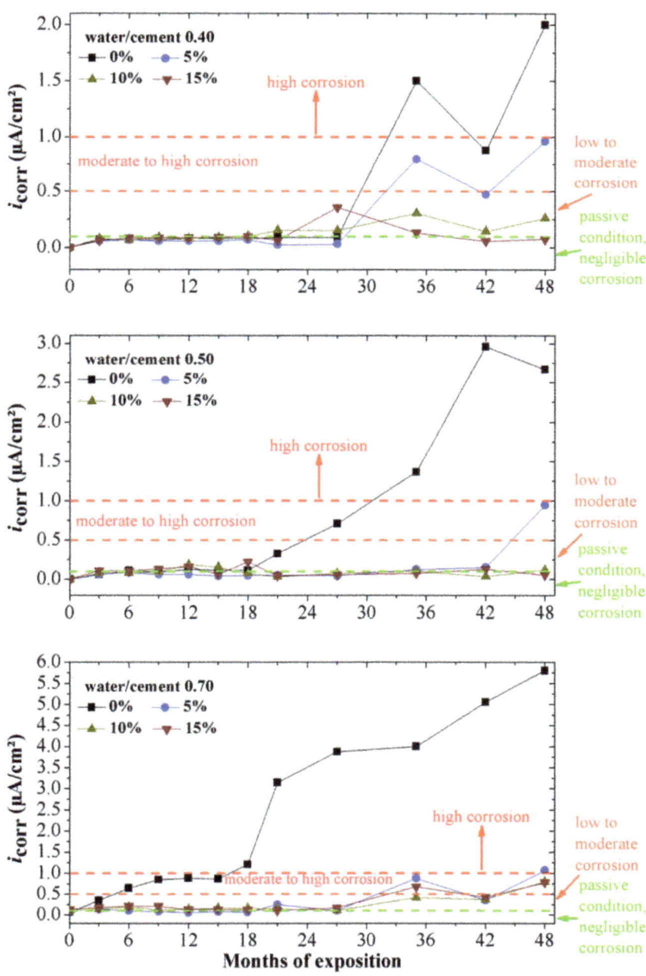

Figure 10.5. Corrosion current density, i_{corr} of reinforced samples exposed to the marine breeze.

lower w/c ratios and high for w/c higher. The corrosion current density for mixtures with a w/c ratio of 0.40, 0.50, and 0.70 reached i_{corr} values of 2, 3, and 6 $\mu A\ cm^{-2}$, which correspond to 20, 30, and 60 $\mu m\ year^{-1}$, respectively [47]. This behavior is caused by the excess water in mixtures with higher w/c ratio which cause more pores than those with lower water content. However, the replacement of cement by RGP seems to change such behavior because mixtures in which cement was replaced for RGP showed a corrosion current density in the region of moderate or negligible corrosion, even if the corrosion onset was established. Such behavior may be attributed to a change in the pore microstructure because when fine ground recycled glass is exposed to an aqueous mixture, it undergoes chemical transformations at its surface compared to the bulk of the glass [48] and modifies the interfacial transition zone (ITZ). In fact, when these particles are added to the concrete, the pH play an important role in the glass corrosion. The alkaline environment in concrete generates an elderly degradation of the glass surface due to surface reaction control mechanisms based on the affinity of reacting species to the surface. Therefore, during the pozzolanic reactivity, cations of Si^{+} from the glass surface may bond with calcium hydroxide to form calcium silicate hydrate. Such bonds decrease the separation the ITZ between glass particles and the cement paste. Calcium hydroxide normally grows as a hexagonal-platelet morphology as a normal curing is carried out. It has been observed that in normal curing the ITZ could be weaker, attributed probably to low hydration, as is observed for a w/c ratio of 0.70 for 15% of RGP. On the other hand, steam curing and microwave radiation heating to 100 °C can increase the compressive strength because more cement particles can be hydrated and the gasification of water and air expansion generates capillary pores, respectively [49, 50]. However, the pozzolanic behavior reactions require that the powder glass size particles should be less than 38 μm [51]. Li and Li [52] proposed a concrete pore classification based on a range of pore diameters, gel pores (less than 10 nm), transitional pores (between 10 and 100 nm) capillary pores (between 100 and 1000 nm) and macropores (more than 1000 nm). Based on such classification, Peng *et al* [4] reported that in cement pastes with RGP, after 90 days of curing, the proportion of capillary and macropores decreased, while the gel pore fraction increased as the replacement of cement by RGP increased. They attributed this behavior to the formation of additional gels of C-S-H (calcium silicate hydrate). The improving of the ITZ and the refinement of the pore structure was also reported by some authors [53, 54]. Then, the findings in this research may be attributed to the densification of IZT and the micropore structure, even if it seems that the replacement of cement by 5%, 10% and 15% RGP may not be optimal because the chloride penetration and concrete carbonation had significant progress, after 40 months of exposure to a marine environment.

10.5 Conclusions

As observed from findings in this report and data from the literature, RGP represents an alternative to the use of raw material in the construction sector and may contribute to the mitigation of global warming, as well as impacting properties such as resistance, strength, and workability. Also, it was proved that RGP has a beneficial resistance to chloride penetration and corrosion to marine exposure environments, because it refines the pore structure of concrete. After a 48-month exposure period, long-term performance studies demonstrate favorable results in terms of both compressive strength and durability.

Further studies must be conducted on the effects of RGP as a replacement of cement or natural sand at different levels and its effects on the pore structure, when local materials are used, aiming to find the optimal conditions to enhance fresh and hardened properties of concrete and mortars and improve the functionality and sustainability of the construction of infrastructure in the future.

Finally, for the successful use of RGP as aggregate on concrete and mortars for domestic and industrial applications it is important to characterize the physical and chemical properties of recycled glass waste, since these are important factors that influence the quality, resistance, and durability of concrete.

References

[1] Yu Y, Chen X, Gao W, Wu D and Castel A 2019 Impact of atmospheric marine environment on cementitious materials *Corros. Sci.* **148** 366–78

[2] Shi X, Xie N, Fortune K and Gong J 2012 Durability of steel reinforced concrete in chloride environments: an overview *Constr. Build. Mater.* **30** 125–38

[3] Islam G S, Rahman M and Kazi N 2017 Waste glass powder as partial replacement of cement for sustainable concrete practice *Int. J. Sustain. Built Environ.* **6** 37–44

[4] Peng L, Zhao Y, Ban J, Wang Y, Shen P, Lu J X and Poon C S 2023 Enhancing the corrosion resistance of recycled aggregate concrete by incorporating waste glass powder *Cem. Concr. Compos.* **137** 104909

[5] Tamanna N and Tuladhar R 2020 Sustainable use of recycled glass powder as cement replacement in concrete *Open Waste Manag. J.* **13** 1–13

[6] Tan K H and Du H 2013 Use of waste glass as sand in mortar: part I—fresh, mechanical and durability properties *Cem. Concr. Compos.* **35** 109–17

[7] Miller S A, Horvath A and Monteiro P J M 2018 Impacts of booming concrete production on water resources worldwide *Nat. Sustain.* **1** 69–76

[8] Hoil-Canul E, Maldonado-López L, Díaz-Ballote L and Casanova-Calam G 2022 Corrosion of rebars in recycled glass concrete under tropical marine environment *Mater. Werks.* **53** 1410–20

[9] Xie J, Wu Z, Zhang X, Hu X and Shi C 2023 Trends and developments in low-heat portland cement and concrete: a review *Constr. Build. Mater.* **392** 131535

[10] Schneider M, Romer M, Tschudin M and Bolio H 2011 Sustainable cement production—present and future *Cem. Concr. Res.* **41** 642–50

[11] Jani Y and Hogland 2014 Waste glass in the production of cement and concrete—a review *J. Environ. Chem. Eng.* **2** 1767–75

[12] Ahmed K S and Rana L R 2023 Fresh and hardened properties of concrete containing recycled waste glass: a review *J. Buil. Eng.* **70** 106327

[13] EU's glass value chain confirms glass collection rate steady progress at 80.1%. FEVE 2023 https://feve.org/eu-glass-value-chain-80-collection-rate/ (accessed 19 November 2023)

[14] Recovery magazine—recycling technology worldwide n.d. https://recovery-worldwide.com/en/index.html (accessed 20 November 2023)
[15] Home—the American School Foundation n.d. https://asf.edu.mx/ (accessed 21 November 2023)
[16] Standard Specification for Mortar for Unit Masonry n.d. https://astm.org/c0270-19ae01.html (accessed 19 November 2023)
[17] Drzymała T, Zegardło B and Tofilo P 2020 Properties of concrete containing recycled glass aggregates produced of exploded lighting materials *Materials* **13** 226
[18] Matos A M and Sousa-Coutinho J 2012 Durability of mortar using waste glass powder as cement replacement *Constr. Build. Mater.* **36** 205–15
[19] Halbiniak J and Major M 2019 The use of waste glass for cement production *IOP Conf. Ser.: Mater. Sci. Eng.* **585** 012008
[20] Huseien G F, Hamzah H K, Sam A R M, Khalid N H A, Shah K W, Deogrescu D P and Mirza J 2020 Alkali-activated mortars blended with glass bottle waste nano powder: environmental benefit and sustainability *J. Clean. Prod.* **243** 118636
[21] Long W J, Zhang X, Xie J, Kou S, Luo Q, Wei J and Feng G L 2022 Recycling of waste cathode ray tube glass through fly ash-slag geopolymer mortar *Constr. Build. Mater.* **322** 126454
[22] Sikora P, Horszczaruk E and Rucinska T 2015 The effect of nanosilica and titanium dioxide on the mechanical and self-cleaning properties of waste-glass cement mortar *Procedia Eng.* **108** 146–53
[23] Zhang W, Li S, Song L, Sheng Y, Xiao J and Zhang T 2023 Studying the effects of varied dosages and grinding times on the mechanical properties of mortar *Sustain. Sci. Pract. Policy* **15** 5936
[24] Matos A M and Sousa-Coutinho J 2016 Waste glass powder in cement: macro and micro scale study *Adv. Cem. Res* **28** 423–32
[25] Khmiri A, Chaabouni M and Samet B 2013 Chemical behaviour of ground waste glass when used as partial cement replacement in mortars *Constr. Build. Mater.* **44** 74–80
[26] Wang Y, Cao Y, Zhang P and Ma A Y 2020 Effective utilization of waste glass as cementitious powder and construction sand in mortar *Materials* **13** 707
[27] Qaidi S, Najm H M, Abed S M, Özkılıç Y O, Al Dughaishi H, Alosta M *et al* 2022 Concrete containing waste glass as an environmentally friendly aggregate: a review on fresh and mechanical characteristics *Materials* **15** 6222
[28] Harrison E, Berenjian A and Seifan M 2020 Recycling of waste glass as aggregate in cement-based materials *Environ. Sci. Ecotechnol.* **4** 100064
[29] Tamanna N, Tuladhar R and Sivakugan N 2020 Performance of recycled waste glass sand as partial replacement of sand in concrete *Constr. Build. Mater.* **239** 117804
[30] Du H and Tan K H 2017 Properties of high volume glass powder concrete *Cem. Concr. Compos.* **75** 22–9
[31] Bignozzi M C, Saccani A, Barbieri L and Lancellotti I 2015 Glass waste as supplementary cementing materials: the effects of glass chemical composition *Cem. Concr. Compos.* **55** 45–52
[32] Du H and Tan K H 2014 Waste glass powder as cement replacement in concrete *J. Adv. Concr. Technol.* **12** 468–77
[33] Chen W, Dong S, Liu Y, Liang Y and Skoczylas F 2022 Effect of waste glass as fine aggregate on properties of mortar *Materials* **15** 8499
[34] Bui H, Boutouil M, Levacher D and Sebaibi N 2021 Evaluation of the influence of accelerated carbonation on the microstructure and mechanical characteristics of coconut fibre-reinforced cementitious matrix *J. Build. Eng.* **39** 102269

[35] Bui H, Delattre F and Levacher D 2023 Experimental methods to evaluate the carbonation degree in concrete—state of the art review *NATO Adv. Sci. Inst. Ser. E Appl. Sci.* **13** 2533
[36] Chang C F and Chen J W 2006 The experimental investigation of concrete carbonation depth *Cem. Concr. Res.* **36** 1760–7
[37] Trocónis-Rincón O, Romero-Carruyo A and Andrade C 2000 *Manual for Inspecting, Evaluating and Diagnosing Corrosion in Reinforced Concrete Structures* (Durar Network Maracaibo)
[38] Broomfield J P 1994 Corrosion rate measurements in reinforced concrete structures by a linear polarization device *SP-151: Concrete Bridges In Aggressive Environments* (American Concrete Institute)
[39] Standard Test Method for Density, Absorption, and Voids in Hardened Concrete n.d. https://astm.org/c0642-21.html (Accessed 19 November 2023)
[40] Yu M-Y, Lee J-Y and Chung C-W 2010 The application of various indicators for the estimation of carbonation and h of cement based materials *J. Test. Eval.* **38** 534–40
[41] García-González C A, Hidalgo A, Andrade C, Cruz Alonso M, Fraile J, López-Periago A M *et al* 2006 Modification of composition and microstructure of portland cement pastes as a result of natural and supercritical carbonation procedures *Ind. Eng. Chem. Res.* **45** 4985–92
[42] Standard Test Methods for Water-Soluble Chlorides Present as Admixtures in Graded Aggregate Road Mixes (Withdrawn 2018) n.d. https://astm.org/d1411-09.html (Accessed 19 November 2023)
[43] Castro-Borges P, Balancán-Zapata M and Zozaya-Ortiz A 2017 Electrochemical meaning of cumulative corrosion rate for reinforced concrete in a tropical natural marine environment *Adv. Mater. Sci. Eng.* **2017** 6973605
[44] ISO 9223:2012. ISO 2022 https://iso.org/standard/53499.html (accessed 19 November 2023)
[45] Liu J, Tang K, Qiu Q, Pan D, Lei Z and Xing F 2014 Experimental investigation on pore structure characterization of concrete exposed to water and chlorides *Materials* **7** 6646–59
[46] Simonsson I, Sögaard C, Rambaran M and Abbas Z 2018 The specific co-ion effect on gelling and surface charging of silica nanoparticles: speculation or reality *Colloids Surf. A Physicochem. Eng. Asp.* **559** 334–41
[47] Fontana M G and Greene N D 1967 *Corrosion Engineering* (McGraw-Hill)
[48] Conradt R 2008 Chemical durability of oxide glasses in aqueous solutions: a review *J. Am. Ceram. Soc.* **91** 728–35
[49] Kong Y, Wang P, Liu S, Zhao G and Peng Y 2016 SEM analysis of the interfacial transition zone between cement-glass powder paste and aggregate of mortar under microwave curing *Materials* **9** 733
[50] Scrivener K L, Crumbie A K and Laugesen P 2004 The interfacial transition zone (ITZ) between cement paste and aggregate in concrete *Interface Sci.* **12** 411–21
[51] Shao Y, Lefort T, Moras S and Rodriguez D 2000 Studies on concrete containing ground waste glass *Cem. Concr. Res.* **30** 91–100
[52] Li Y and Li 2014 Capillary tension theory for prediction of early autogenous shrinkage of self-consolidating concrete *Constr. Build. Mater.* **53** 511–6
[53] Lu J X, Duan Z H and Poon C S 2017 Combined use of waste glass powder and cullet in architectural mortar *Cem. Concr. Compos.* **82** 34–44
[54] Corinaldesi V, Gnappi G, Moriconi G and Montenero A J W M 2005 Reuse of ground waste glass as aggregate for mortars *Waste Manage* **25** 197–201

www.ingramcontent.com/pod-product-compliance
Ingram Content Group UK Ltd.
Pitfield, Milton Keynes, MK11 3LW, UK
UKHW051856100926
8556IPUK00005B/652

9 780750 359